Industrial Noise Control

INDUSTRIAL NOISE CONTROL

Bruce Fader

Senior Consultant
Donley, Miller & Nowikas, Inc.

A WILEY-INTERSCIENCE PUBLICATION

JOHN WILEY & SONS
New York • Chichester • Brisbane • Toronto

Library of Congress Cataloging in Publication Data:

Fader, Bruce.
  Industrial noise control.

  Includes index.
  1. Noise control.   I. Title.

TD892.F26   620.2′3   81-2158
ISBN 0-471-06007-0   AACR2

Printed in the United States of America

10 9 8 7 6 5 4 3 2 1

PREFACE

"If you don't have problems," an old boss of mine once said, "you don't have a job!" He was a kindly man, though, and so am I. Probably your boss said to you, one day, "You are in charge of noise control." (Mine did.) Many people have helped me live up to that responsibility. Very humbly, I'd like to help you with it now.

Many of the standard and classical texts on acoustics are time-consuming and almost arcane when you have a real problem to solve in a hurry. And yet some insight about how noise is created and propagated is really indispensable in calling forth noise control measures. There are few fields where you can spend so much money and accomplish so little if you do not have the basic ideas in hand.

Giants have preceded us here in acoustics, as in other fields of science. *They* are not acknowledged here. This is a shirt-sleeve view of the subject.

Still, I owe a debt to some people who—while not giants, perhaps—were well above average height. My thanks go especially to the late Ed Shippee (a most kindly guy), Rockie Rocca of Barry Corp., who knows more about vibration isolation than I shall ever learn, and Harold S. Lance of American Air Filter who supplied information and checked fine points about mufflers and silencers.

Thanks, too, to George Diehl for his painstaking examination of the manuscript and invaluable help and Andrew Borloz for his illustrations.

Perhaps the most kindly guy of all was Dick Guernsey of Cedar Knolls Acoustical Laboratory, who waded through all this while it was still badly typed. He argued—doggedly, but with all good humor—about every issue until it was brought to the ground.

My most personal thanks go to W. Ranger Farrell. He was the godsend who helped me begin to make some sense of it all shortly after that day when the boss said, "You are in charge of noise control."

BRUCE FADER

East Hanover, New Jersey
May 1981

CONTENTS

Industrial Noise Control

INTRODUCTION

This is not the first book ever written on noise control and undoubtedly it will not be the last. Why, then, has someone taken the trouble to write yet another book?

For one simple and compelling reason: the available books seem to miss a particular mark. This book supposes that you are not afraid of doing some thinking and calculations and that you earnestly want or need to control industrial noise to the satisfaction of the Occupational Safety and Health Administration (OSHA). It proposes to help you do that in the simplest and most direct way.

Most books on the subject of acoustics look inward, viewing acoustics as a discipline in its own right. And so they should. It is not a meager subject. There are puzzles, even mysteries, in it—but the triumph of having already solved some of them is part of its history. This book looks outward. It has been written with the idea that your need to cope with noise problems is immediate. It proposes to show you how to do that.

On the other hand, this book does not thumb its nose at the theory, insights, and differential equations in the more learned texts. In fact, it may lead you to look them up and study them.

But there are no differential equations here. You'll find only the working methods of the people who earn their living in industrial noise control. This book has a shirt-sleeve approach. It works.

1

WHAT IS SOUND?

All good books on acoustics and noise control start with this question. In a way, of course, the question What is sound? is silly. Sound is commonplace. It is what we hear—continually—and have heard all our lives. It is sensed by the ear. The ear works because the eardrum is made to vibrate by the air in contact with it when there is sound to be heard.

If the eardrum vibrates because it is in contact with the air and there is sound in the air, should we conclude that sound is a vibration in the air? As a starting place, yes. It is a peculiar sort of vibration, though, and one which is not like the vibration of a pendulum.

Except for the case of the air in an organ pipe or a seashell (or their counterparts) sound is a *forced* vibration in the air. The vibrational energy we call sound is just passing through. Sound doesn't loaf along, either! The speed of sound is 1130 fps. That speed is emphasized here so that it will be easy to find. You will have use for it later.

Unless there is a continuing source pumping power into the air, the sound will cease promptly. So the air doesn't vibrate like the old-fashioned spring on the screen door or a pendulum. Normally, the air is not part of a resonant system; it is only conveying vibrational energy that has been created elsewhere. It brings some of this energy to our ears when it is forced to.

The speed of sound (greater than that of the average rifle bullet) is interesting. Strictly, it should be said that at one atmosphere and 68°F the speed of sound is 1128 fps. No doubt you correctly infer that for other temperatures, other pressures, and other gases, sound's speed is different. Indeed, this is so.

THE LOCAL SPEED LIMIT

The local speed limit for sound in a gas is close to the average velocity of the molecules in that gas. There are many ways of defining that average velocity. They involve the temperature, molecular weight of the gas, and a ratio of heat capacity of the gas at constant pressure and constant volume. (See Table 1.) There is no need to go deeply into the physical theory.

Table 1 *Speed of Sound Data*

Gas	Molecular Weight (Daltons)	γ Ratio of Specific Heats
Air	29.0	1.40
Ammonia	17.0	1.31
Argon	39.9	1.67
Carbon dioxide	44.0	1.30
Carbon monoxide	28.0	1.40
Chlorine	70.9	1.37
Ethane	30.1	1.19
Ethylene	28.1	1.25
Helium	4.0	1.66
Hydrogen	2.0	1.41
Hydrogen sulfide	34.1	1.33
Methane	16.0	1.31
Nitrogen	28.0	1.40
Oxygen	32.0	1.40
Propane	44.1	1.14
Sulfur dioxide	64.1	1.24
Water vapor	18.0	1.33

The general idea is helpful in understanding some sorts of noise control methods. The argument runs this way: Air, for example, consists of a mixture of molecules. They have an average molecular weight of 29 Daltons. The energy contained in the air is expressed as $mv^2/2$, where m is the molecular weight and v is the velocity of the molecules. But the energy contained in the air can also be expressed as a function of temperature. Since the molecular weight can't change, the average molecular velocity must change with temperature. For air at room temperature, that velocity is about 1128 fps. This is also the speed at which any impulse—like the "vibration" we call sound—can be conveyed through the air.

As the temperature of the air increases, the speed of sound also increases but only as the square root of the absolute temperature. In English units, absolute temperature is given in degrees Rankine. To find the absolute temperature add 460 to the temperature in degrees Fahrenheit. Figure 1 will spare you any calculations as long as you are working with air.

For gases other than air, the speed of sound can be calculated by

$$c = 223 \sqrt{\gamma T/m}$$

where c = speed of sound (fps)
   γ = ratio of specific heats (no units)
   T = absolute temperature (°R)
   m = molecular weight (Daltons)

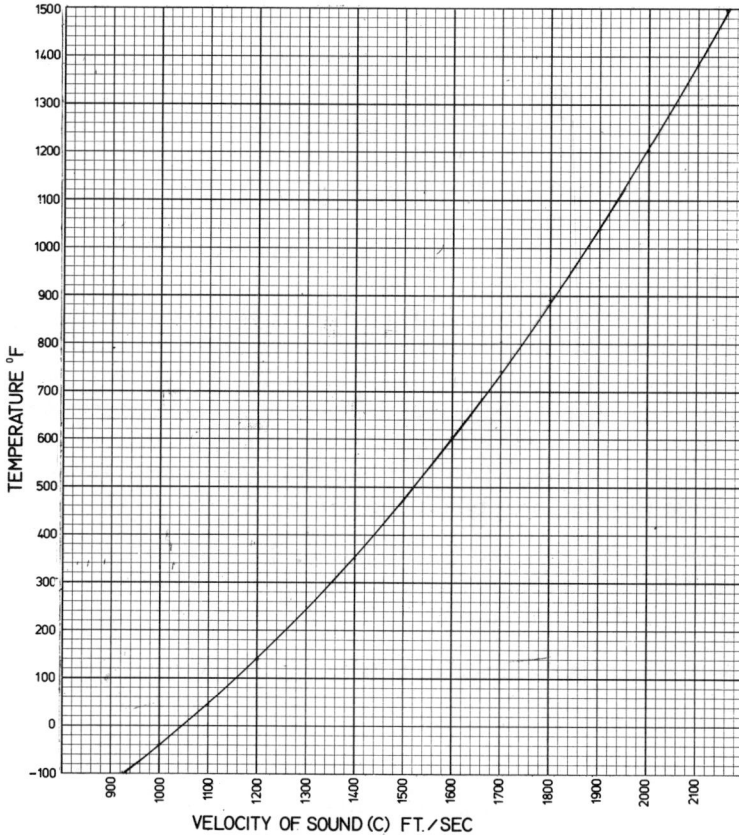

Figure 1　The speed of sound in air can be read directly from this curve.

Thus you could estimate that sound would travel at about 1050 fps in pure argon, or 3300 fps in pure helium, at room temperature.

Do these interesting insights about sound traveling through gases have any practical applications? Consider, for example, that divers working at extreme depths usually breathe a mixture of oxygen and helium. The speed of sound in this mixed gas (average molecular weight of 9.6) can be estimated as about 2060 fps. There is a real problem talking to these divers as they work. They all sound like Donald Duck! The phenomenon is known as "helium speech."

What would happen if a mixture of oxygen-helium-argon, such that the average molecular weight was 29, were to be substituted for the oxygen-helium mix for these divers? You have just solved your first problem in acoustics. Remember, air has a molecular weight of 29. The first requirement would be to prove that it is a safe mixture to use, and you go on from there.

You might also—quite logically—have suspected that the pressure under which the divers work had something to do with helium speech. Actually pressure is usually unimportant to the speed of sound. The sound power that gases convey is transmitted by molecules hitting other molecules. All that high pressure can do is to reduce the distance a molecule has to travel, on the average, before it hits another molecule.

The collision and transfer of kinetic energy is supposed to be elastic and instantaneous. In gases the molecules are always so far apart that in any ordinary noise control work you will never have to make any correction for pressure in the speed of sound. For gases compressed until they are nearing the liquid state, the speed of sound increases sharply with increasing pressure (this idea is explained later in the section on liquids and solids).

Do not conclude that there are no important practical effects connected with the speed of sound. Many acoustical devices depend critically on the speed of sound. In a later chapter you'll see that the muffler on your car is one of them. Had it been designed on the basis that the speed of sound is, as everyone knows, 1130 fps, it wouldn't work very well. Automobile exhaust is a mixture of nitrogen, carbon dioxide, carbon monoxide, and water vapor, and its temperature at the muffler is typically 900°F.

WORKED EXAMPLE   What speed would you assign to sound in the muffler? Assume the exhaust gas composition is carbon dioxide, 15%; carbon monoxide, 1%; water vapor, 8%; and nitrogen, 76%.

First find the average molecular weight and specific heat ratio (γ):

Component	Fraction of Exhaust	Molecular Weight	γ	Fraction of: Molecular Weight	γ
Carbon dioxide	0.15	44.0	1.30	6.60	0.20
Carbon monoxide	0.01	28.0	1.40	0.28	0.01
Water vapor	0.08	18.0	1.33	1.44	0.11
Nitrogen	0.76	28.0	1.40	21.28	1.06
Average molecular weight				29.6	
Average γ					1.38

The temperature of 900°F must be converted to absolute:

$$900°F + 460 = 1360°R$$

The speed of sound should be

$$c = 223\sqrt{\gamma T/m} = 223\sqrt{1.38 \times 1360/29.6}$$

$$c = \text{about 1780 fps}$$

THE EVER-EXPANDING BUBBLE

Why and how does the speed of the gas molecules set the speed of sound? Imagine what is happening in a small part of the air in the room you're in right now. In your mind, block out a cubic foot of the air. Now imagine that the ventilation has been turned off and that there are no drafts. Is the air still? Yes, in the sense that if you dropped a feather or piece of lint, it would fall straight down. If you were smoking, however, you would notice that the smoke gradually diffused, which is fairly convincing evidence that the air molecules are flying randomly in all directions.

Now suppose you leave five sides of that box (your imagined cubic foot of air) stationary and slowly move one side in about an inch. Can you suppose that the air near the wall is compressed while the air in the rest of the cube remains unchanged?

Suppose you moved that wall in faster, and still faster. When *would* the air near it begin to be compressed without the air in the rest of the cube being disturbed? Let's be clear: given even a moment, all the air in the cube would be at the same pressure, of course. What we are talking about here is a transient effect. The answer should begin to be apparent. The air molecules will begin to become crowded near the moving wall when its speed begins to approach the speed with which the air molecules can escape. *That* speed is the speed of the molecules and hence also the speed of sound.

An analogy to this little mental experiment exists in the real world and you are familiar with it. When jets fly at or above the speed of sound they create the well-known sonic boom in just this way.

While we have our mental lab equipment out, let's make another experiment. Put the wall of the cube back where it was originally and suppose now that the walls won't move and won't transmit sound. With an imaginary drill, drill an imaginary hole in the top of the box and drop in a lit firecracker. Be sure to stand well to one side of the cube (this is not a safety precaution—you can't be hurt by the firecracker, it's only imaginary!)—standing well to one side is part of the experiment.

What happens next? The firecracker goes off and you hear the bang. How? The explanation lies with those air molecules. This time they were given a really sudden shove by the exploding firecracker. They were pushed away in all directions. They crowded the molecules in front of them, passing along the impulse, and *those* molecules did the same. In fact, an instant after the explosion there

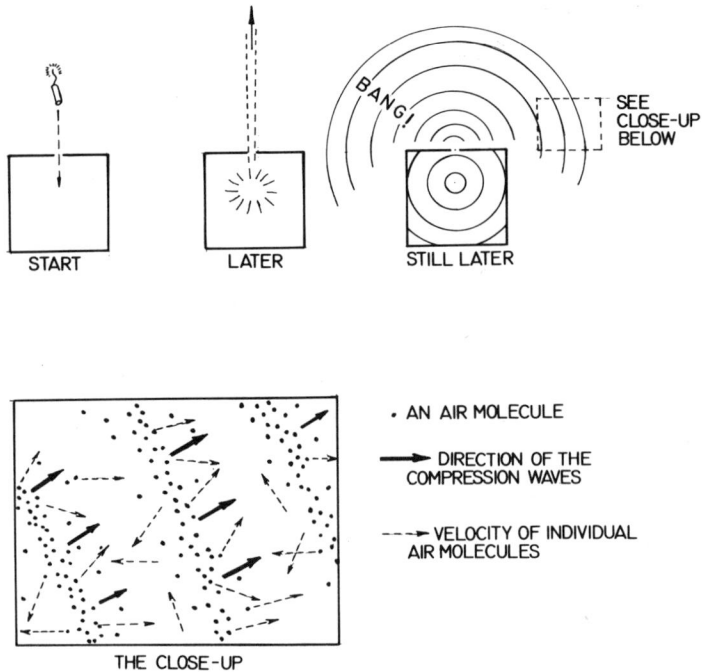

Figure 2 A mental experiment with a firecracker and a box provides insight into how sound propagates in air. The flash of light that comes through the hole in the box is a narrow beam. The sound spreads out in concentric spheres centered on the hole. This sort of spreading is a consequence of the random directions taken by individual air molecules.

was a partial vacuum where the firecracker had been and a relatively crowded spherical shell surrounding it.

The first molecules to have been moved by the explosion had already passed their outward velocity along to those in front of them. Because of the crowding in that outward-bound shell, the original molecules began to rebound into the partial vacuum behind them. Soon the spot where the firecracker had been was crowded again—though not so crowded as in the instant following the explosion—and the whole process began again. Thus several of these concentric spherical shells radiated from the explosion point.

DIFFRACTION—VIRTUAL SOURCES

The reason you can hear the bang while standing to one side of the hole is connected with a curious thing that happens when each shell reaches the hole. To best understand, think of one of the unsuspecting air molecules just outside the hole when one of those compression shells arrives from inside the box. That molecule will be hit by one or more of the jostling molecules of the expanding shell and, while they rebound, it will be pushed off in *some* direction. About the only generalization that applies to *that* direction is that probably it will be away from the hole rather than toward it.

The process is called diffraction—a kind of bending of the path of these compression waves when they pass an edge or an obstacle.

Diffraction can be analyzed and studied at length. It won't be here. In working at noise control, however, there is one generalization about diffraction that is worth remembering: The longer the wavelength, the more easily the sound energy bends when it comes to an edge or an opening.

Now, as an observer, you can see that it is perfectly possible that the string of compression waves that make up the sound of the explosion in the box could reach you even though you were to one side of the open hole. In one sense, in fact, it was the hole and not the firecracker that was the source of the noise.

One more mental experiment helps clear up that idea and illustrates why diffraction is important in working with noise. Using some quick-drying imaginary black paint, paint the whole box so that the only way light can get out is through the hole. Now drop in another imaginary lit firecracker, watch the ceiling, and listen! This time you will see a fairly small patch of the ceiling illuminated by the flash of the exploding firecracker and hear the report so soon after the flash that they seem simultaneous.

The light was reflected from the ceiling. So was some of the sound. But some of the sound, as explained previously, reached your ear directly from the hole even though it was not in your line of sight. The light traveled in a slender beam. The sound was diffracted* from the hole and reflected from the ceiling, and the hole was its apparent (or virtual) source.

Viewed in this way, sound can be seen to be a very special kind of vibration in the air. Although "vibration" serves to characterize many acoustic effects of sound, the air doesn't vibrate in the sense that a spring-mass system or a bowl full of jelly does.

LIQUIDS AND SOLIDS

You can say that sound exists and travels in both liquids and solids. In fact, sonar and submarine detection and underwater communications have opened up a big field of underwater acoustics. Geologists and seismologists depend on sound waves in solids to learn about subsurface features. Nondestructive testing makes use of a few acoustic techniques that depend on how sound travels through metal parts and so forth.

The speed of sound is much higher in both liquids and solids than in gases. This does not mean that the molecules of solids and liquids are traveling faster than those in air. Consider that 1 ft^3 of water turns into about 1000 ft^3 of steam when you boil it. Think how much more densely the water molecules are packed

---

*Light is diffracted, too, but the effect is minute compared to the effect in the case of sound.

in the liquid form. There is not much space between them. In a solid crystal, the molecules are as close as they can get to each other (Almost! Crystals can be strained or compressed slightly.)

Now when a mechanical impulse is introduced into a liquid or solid, its transmission must be rapid, though not instantaneous. You've probably seen the toy (Figure 3) that consists of a frame supporting five steel balls that touch each other. When you lift a ball and let it swing into one end, the ball at the other end seems to pop out at the same instant. That same mechanism is at work in transmitting sound in liquids and solids.

Another difference between the three states of matter is that while sound really does radiate in spherical shells in gases, it is more readily concentrated into a beam in liquids, and it frequently propagates almost entirely in preferred directions in some solids. A case in point: a few years back a New York office building was getting constant tenant complaints about a rumbling. It was found to be caused by New York Central trains under Park Avenue. The building in question was three blocks east and *none of the intervening buildings were having any trouble!* Some peculiarity of the bedrock, perhaps, channeled sound to that one building.

Is "vibration" a good word to describe sound in liquids and solids? Yes, but again you should distinguish between waves that are merely on their way through the medium and resonant vibration. It is a matter of convenience, really, for people who work at noise control. Be careful to use words like "ringing" or "resonant" to describe solid parts that are sources of bell-like noise when they are struck. Thus it will be clear that when you say the rail is vibrating because of a train a mile away, you really are talking about sound in the rail.

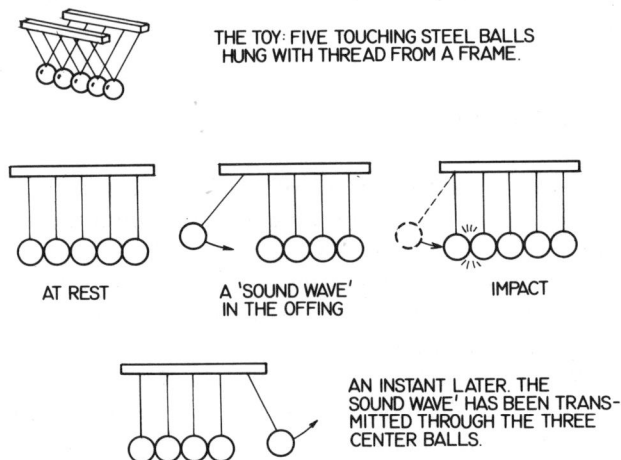

THE TOY: FIVE TOUCHING STEEL BALLS HUNG WITH THREAD FROM A FRAME.

AT REST   A 'SOUND WAVE' IN THE OFFING   IMPACT

AN INSTANT LATER. THE 'SOUND WAVE' HAS BEEN TRANSMITTED THROUGH THE THREE CENTER BALLS.

Figure 3 The mechanism of sound (vibration) transmission in solids makes use of the closely packed and well-ordered arrangement of the atoms.

TRICKS SOUND PLAYS

Whispering galleries were a toy of the Renaissance. There is one in the Capitol in Washington. Another you may have a chance to try personally is in Grand Central Station in New York. Between the upper and lower levels there are a pair of east-west ramps that descend to a central platform where another north-bound ramp leads to the final lower elevation. The square intersection has masonry arches in both views and they are connected with a smooth curved transitional surface. The square is about 35 ft on an edge. An architect ex-plained it to a friend, and they tried it during a rush hour.

While literally hundreds of people walked by to catch trains, the architect stood facing one corner and his friend faced the diagonally opposite corner about 50 ft away. They talked to each other in easy conversational tones—and none of the passers-by could eavesdrop beyond a range of a few feet (though they may have wondered about those strange guys talking to the corners).

What was happening, of course, was that the curved ceiling surface focused the sound emanating from one point to the other point. The architect and his friend were each standing at one of those focal points.

Has this anything to do with industrial noise control? Try a plant with a roof made of 80 ft clear span arches (the trusses may be 25 or 30 ft deep at the center). Guess where the noisiest equipment happens to be located? At the focus, of course!

Another little trick that noise plays sometimes has to do with *standing waves*. You have seen them on the surface of a cup of coffee served in a train—little ripples that appear to stand still. They are caused by the spatial relationship between the energy source and the reflecting surfaces.

A peak arises at all those points where the wavefront reflected from a surface *always* arrives at the same instant as the wavefront directly from the source (or from another reflecting surface). In order for standing waves to be a problem (or, at least to be interesting) the noise source must be producing a steady stream of waves of exactly the same length (otherwise, they could not consis-tently reinforce one another at the same point in space).

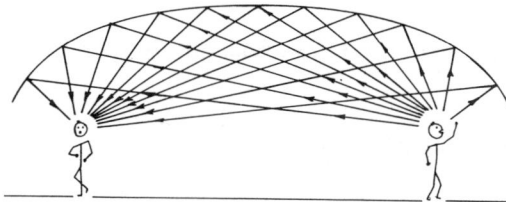

THE WHISPERING GALLERY

Figure 4 One type of whispering gallery depends on surfaces that focus the sound waves. In this case, the ceiling is approximately elliptical.

By wavelength, of course, we describe the distance between the successive spherical shells that radiated from a noise source. In an ideal case the source would be very small like the hole in the box of the firecracker experiment. The waves would pass any point (your ear or a microphone) at equal intervals of time. The reciprocal of that time would be the frequency at which they passed the point.

A simple, obvious example would be that if $1/100$ sec elapsed between successive waves, the waves would have a frequency of 100 per second. It is also very easy to find the wavelength if the speed of sound is known:

$$\lambda = 1130/F$$

where λ = the wavelength of sound (ft)
            1130 = the speed of sound (for air at
                    room temperature) (fps)
            F = the frequency of the sound (Hz)

Do you have a transformer at your plant? If it is near a good reflecting surface, walk around in the space between the transformer and the reflector and listen critically. You will find some spots where moving your head through a few feet produces a big change in the noise.

Now standing waves are not the most common problem in noise control work, but there is a special case where they can be a real horror. Suppose one of these noise sources that produces a steady stream of waves of exactly the same length is located at the open end of a tube (closed at the far end) that is exactly one quarter wavelength long. Any odd multiple ($1/4$, $3/4$, $5/4$, $7/4$, . . .) of a quarter wavelength does just as well. What happens now? Every time the noise source produces a wavefront, a reflected wavefront is arriving at the source from the tube. In a rare case, this might cause the source to be about four times as loud as it would be without that tuned tube. It is commonplace for it to be twice as loud.

This never happens? In the preceding explanation, we had the source at the open end of the tube. In practice it really doesn't matter where the source is in the tube—even at the "closed end." Now there was an industrial vacuum cleaner with a four-bladed steel plate fan that was mounted to a steel discharge duct that carried the dirt into an open bin. The fan turned up 3200 rpm. How long do you suppose the duct was?

9 ft 3 in.—a perfect organ pipe ($7/4$ wavelength) for the tone produced (213 Hz)

MORE COMMON PROBLEMS

In this chapter we have reviewed what everyone knows about sound and, perhaps, introduced a new idea or two. One mathematical notion was raised—to demonstrate a point—but there has been no heavy emphasis on numbers and calculations.

What must be emphasized is that sound is a wave phenomenon (compression waves, when we deal with it in air), and it has a velocity. Sound waves can be reflected (the industrial vacuum cleaner, the whispering gallery, the standing waves near a transformer). It is not surprising then that sound waves can be absorbed also. You have all been on a beach. There are no strong reflected waves headed back into the ocean when the surf rolls in.

Sound waves can also be blocked. The worst ocean storm causes only a small disturbance behind the big breakwater of a safe harbor. In noise control, this is analogous to putting up a wall to block transmission of noise energy.

2

THE QUALITY
OF SOUND

If you are musical, there is a lot you can gain by paying attention to the things you know about music. For example, you know about pitch—sharp and flat and all that. In fact, everybody knows what pitch is, but in acoustics it is called frequency. It is the frequency with which those compression wavefronts push your eardrum inward (of course followed by partial vacuum in which your eardrum is asked to bulge outward again). The number of times per second this happens is the frequency in hertz. (This used to be called cycles per second, or cps. Today, the unit for frequency has been dedicated to Hertz, one of the fathers of radio. Frequency is very important there.)

If you are muscial, perhaps you took piano lessons. If you did, you have a way to convert the antiseptic numbers of frequency into something you can recall (even if you didn't take piano lessons, you can always pick away at the keyboard to test this idea). The lowest note on an 88-key piano has a frequency of about 27.5 Hz. The highest note is about 4400 Hz. If you took piano lessons, you've noticed that middle C is *not* in the middle of the keyboard. There are about three and one-quarter octaves below it and there are about four octaves above it.

What do you suppose the frequency of middle C is? First, if we shortened the piano keyboard so that it ran from a low A to a high E-flat (chopping off most of the highest octave) middle C would really be in the middle. The frequencies of that low A and the high E-flat are 27.5 and 2500 Hz.

Thus middle C must be the average: $(27.5 + 2500)/2 = 1258$. *Not so!* Middle C (for an A = 440 Hz piano) is 261.63 Hz (see Figure 5).

What went wrong? The ear didn't perceive a 1 Hz change in frequency as well at 2500 Hz as it did at 27.5 Hz, which is natural if you think in terms of the percentage change represented by 1 Hz.

This could be a long story, but it won't be. The story wasn't discovered yesterday. Pythagoras enunciated it 23 centuries ago. He did the first "paper" on how to tune stringed instruments and he knew the consequences of what he was saying. The note that lies one octave above another note on any musical instrument has a frequency twice that of the lower note.

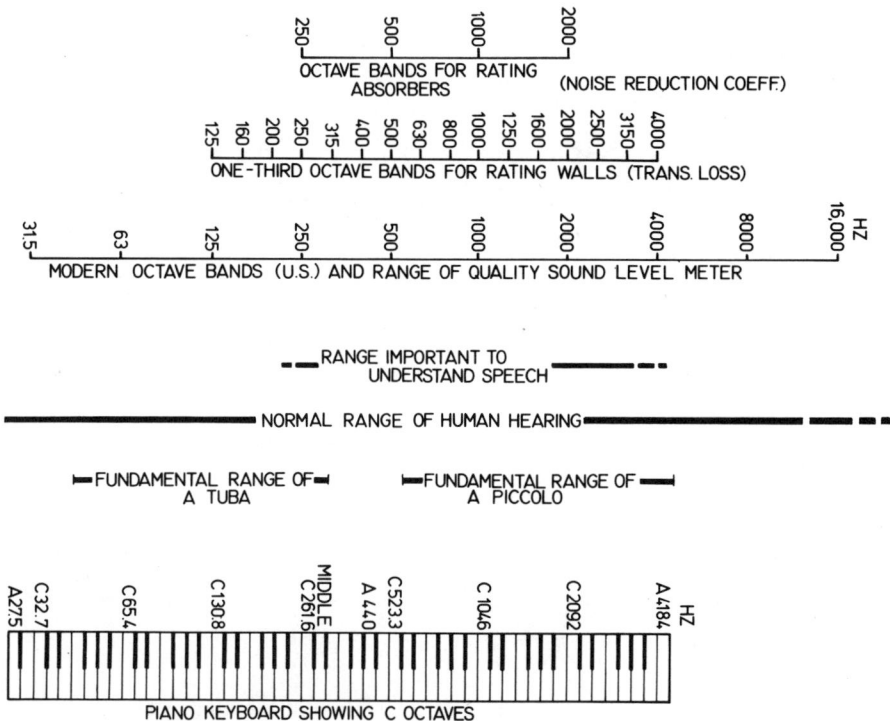

Figure 5 Frequency in relation to the piano keyboard.

This is also true in acoustics. An octave is an octave. If you trust your piano tuner, you can refer to the frequency scales aligned with the piano keyboard (see Figure 5) to verify that middle C is not an arithmetic average of the high and low ends of the piano keyboard.

You might also note that the doubling of frequency for each higher C makes the keyboard—and the usual frequency scales of acoustical instruments—nonlinear. They are, in fact, logarithmic. If you are going to do much work in noise control and acoustics, get to know your local logarithms! Watch for the idea to show up again (and again and again) as you work your way into this little corner of applied science.

THE OTHER SIDE OF THE EARDRUM

We said that we hear sound when the eardrum is caused to vibrate. What happens after that? The eardrum is physically attached to three small bones: the hammer, the anvil, and the stirrup. (Early anatomists missed the third one. No wonder, it is the smallest bone in the human body—about half the size of a grain of rice). The stirrup is attached to an "oval window." This window is the

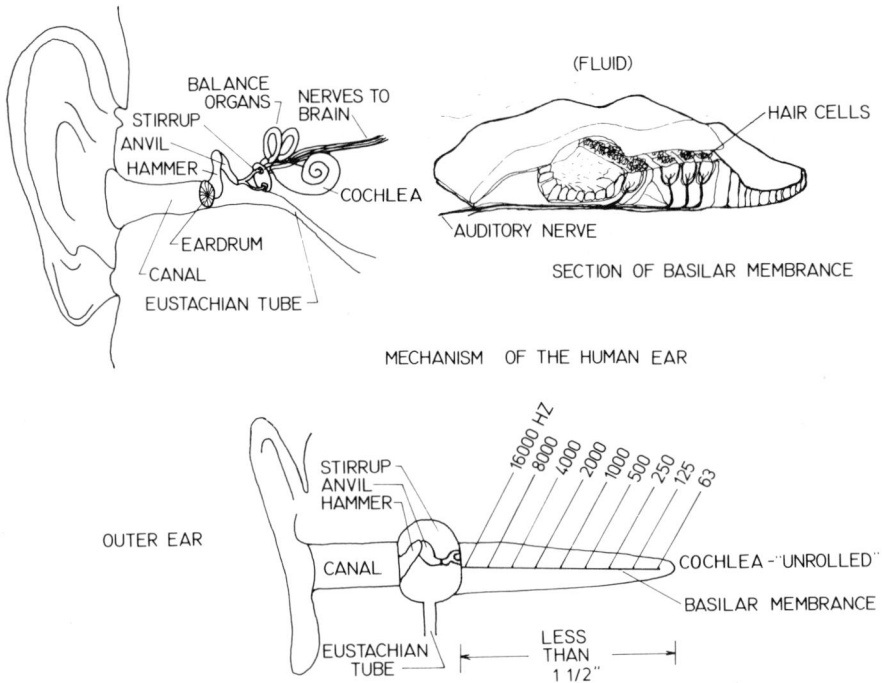

Figure 6 The arrangement and mechanism of the human ear.

broad end of a tapered tube that—if unrolled—would be about 1 ½ in. long (Figure 6). It is coiled up like a snail: hence its name, the cochlea, Latin for snail.

In fact, that high E-flat has a frequency nearer 2490 Hz. The usual approach to finding a "mid frequency" between two extremes is to use

$$f_{mid} = \sqrt{f_{lo}f_{hi}}$$

where f_{mid} = the mid frequency (Hz)
        f_{lo} = the low frequency (Hz)
        f_{hi} = the high frequency (Hz)

From the general region of the cochlea there are two obvious paths that lead away. One of them is the Eustachian tube, which ends in your throat. Its purpose is to allow the air pressure on the outer and inner sides of your eardrum to be equalized. The other path leading away from the cochlea is a nerve bundle that goes to the brain. Neither this book nor the best insights of today's science will explain what is transmitted along that nerve bundle or how we appreciate what it is that we are hearing.

The purpose of this chapter is to take advantage of what you already know about sound to help you understand some other things you will find useful in solving industrial noise problems. If you like long words, the science of exactly how we hear is psychobioacoustics, a branch of physics. Pursuing some of the strange phenomena of this field may not help you solve your plant's noise problem very quickly. Nevertheless, some of the ideas you come to grips with are so fascinating that you'll find yourself wondering about them in your spare time. The ear-brain path is one of the strangest.

Put yourself in this scene: you are in the backyard listening to a football game on a small transistor radio (speaker diameter, 1 ½ in.) Half time comes. Between commercials you are treated to the marching bands performing on the field. You distinctly pick up and follow the steady oom-pah, oom-pah of the tuba in your favorite part of the "Colonel Bogey March."

Now some shocking news. There is no way a 1 ½ in. speaker can generate frequencies below 300 Hz, the highest note it is possible to play on a tuba. (Physical objects smaller than a wavelength of sound radiate it poorly and when they are smaller than one-tenth wavelength, very poorly.) How in the world are you hearing a tuba through a 1½ in. speaker? Refer to Figure 5 and note that it shows the fundamental range of a tuba. For most musical instruments, and most sources of industrial noise—even the ones that produce a tone or note that you can whistle—many multiples of that frequency are produced, too.

The *fundamental* is the note that the tuba player played. It is also the frequency at which sawteeth generate impacts on the stock being sawn. If multiples of that frequency—*harmonics,* to acousticians and *overtones* to musicians—are generated, and they usually are, the harmonics from the tuba could have come through your small speaker. In fact, the tuba, like every musical instrument, has a characteristic set of harmonics. Some of the multiples of the fundamental are small, some large, and some nonexistent. As a matter of fact, it is in just this way, by recognizing the importance of the harmonics, that your brain-ear combination knows that it is a tuba playing and not a violin or clarinet.

Pull yourself back to the backyard and your favorite oom-pah. Your ear-brain never even heard the note the tuba was playing, only its harmonics. It was smart enough to know the note and—from the structure of the spectrum of the harmonics—that it was a tuba. You just recognized the Cheshire cat by his grin alone (and if *you* played tuba in the band, keep in mind that it works the same way with the E-flat euphonium).

HAIR CELLS—AND HEARING LOSS

You need to know what happens in the cochlea. The tapered tube has a membrane running along its length. This is called the basilar membrane. On both sides there is fluid. On the side connected to the oval window, the membrane has cells that have hairs extending out into the fluid.

Under a microscope these projecting "hairs" look more like canoe paddles or stalked spheres than hairs. When the hairs are made to tremble, they send some sort of message to the brain.

One thing that you need to know about the ear is how noise-induced hearing loss is incurred. The hair cells along the basilar membrane are—in a very broad analogy—something like the keys of a piano keyboard. Their location is related to the frequencies they sense. Prolonged exposure to intense noise destroys the hair cells for lower harmonics of the frequencies the intense noise contained. Hair cells cannot be regenerated—the hearing loss is permanent.

You must not push the keyboard analogy too far though. Hair cells near the location of the destroyed cells can sense the frequency associated with the damaged region of the membrane. These are not as perceptive as the destroyed cells were for those frequencies.

THE LIMITS

A healthy youngster who hasn't spent too much time with shotguns, snowmobiles, or discotheques has an amazing range of hearing. In the first place he can hear tones from about 20 (or somewhat less) Hz to perhaps 20,000. Not everyone who reads this book hears the higher frequencies he does. If we all had hearing tests—audiograms—run, we would find a scattered set of curves showing various amounts of hearing loss. In general, we would find the more severe losses among older men who had too much time in noisy places.

Women tend to fare better with respect to high frequency loss, which probably has a lot to do with the perennial argument about the hi-fi between husbands and wives: "George! Will you turn that thing *down*?" Poor George is just trying to hear the fine edge of the trumpet tones. It may require 100 to 1000 times as much sound power (or substantially more) for him to hear them as for her.

But that's not the biggest wonder about the healthy youngster's hearing. If you give him a quiet enough place, he can probably hear his wrist watch ticking even when it is held at arm's length. And he can tolerate shooting a gun without measurable permanent damage (not as a steady diet, though!).

The difference in power between a watch tick and a shotgun blast is a factor of about 10^{16}. In other words, if you could get 10,000,000,000,000,000 watches to tick in unison the sound power would equal that of a shotgun.

You wouldn't hear it that way, though. Just in case you are not impressed with this enormous range of hearing capability, if we are talking about wrist watches (straps removed) measuring $1 \times 1\frac{1}{2}$ in. each, there is only room for about two thirds of those 10,000,000,000,000,000 watches if we lay them out next to each other in the continental ("lower 48") United States.

What sort of hearing equipment did that healthy youngster have to accomodate such a staggering range of sound power? We have already looked at the way the ear is built. Now perhaps we can see why it needs to be built with a complicated linkage of three bones between the eardrum and inner ear. These are, in effect, a volume control. Only when it is very quiet do we develop our best hearing sensitivity. When we are in a noisy place, our ears are less sensitive. It is nature's way of protecting the delicate parts of the system—the hair cells.

THRESHOLD SHIFT

Now when you spend the morning out on the floor of the boiler factory and then discuss a problem with the boss at lunch, do you ever notice that you have trouble hearing every word? You are experiencing a temporary threshold shift. Your ears are slowly and cautiously resetting the volume control for quiet conditions. It's as though they were afraid that those noxious rivet guns and disc grinders were going to start in again at any minute. If you spend too many mornings in the boiler factory, the temporary threshold shift becomes a permanent threshold shift.

It's not a case of the volume control getting stuck. As a matter of fact, your ears have probably adjusted for good sensitivity by then. What has happened is the loss of hair cells. You've burned out your transistors—in the volume control analogy—and there are no replacements!

Another effect of exposure to noise—especially fairly brief exposure to very intense noise—is tinnitus ("TIN-it-tuss") or ringing of the ears. It is a good sign that you have been in a place that needs noise control or where you had better wear hearing protectors if you have to spend much time there. Some of us are blessed with perpetual tinnitus, and it's not pleasant, as those who have a permanent case of it can tell you.

HOW "HI—FI" ARE EARS?

Hardly anybody is "flat plus or minus 1 dB from 20 to 20,000 Hz" as they like to say in the ads for expensive amplifiers.

Let's look at our healthy youngster one last time. He has perfect hearing. How good is that? We've looked at two quantities, frequency and intensity, where that young lad had it all over us. But can he accommodate that range of sound power and still come anywhere near the "flat within 1 dB from ..." requirement? In a word, no.

Fletcher and Munson were two stalwart investigators of this question back in 1936. They have been superseded by a number of others with better instruments, and a blazed trail to follow. They generated the set of curves that appear in Figure 7. What can be said in general about these curves is:

1   In each curve the pure tones in each octave seem as loud (on the average) as those in any other octave.

2   The increment of loudness between each pair of curves is equal. As well as people are able to describe loudness, they would say (on the average) that each curve represents noise "twice as loud" as the curve just below it. (Homework assignment: In a quiet room, tap an ashtray with a pencil and then tap it again to make a noise "twice as loud." This is a "think about it" sort of assignment.)

3   When noise is loud, the response of the ear is more nearly equal at all frequencies than when the noise is quiet.

EQUAL LOUDNESS CONTOURS FOR PURE TONES
BY OCTAVE (FLETCHER-MUNSON CURVES)

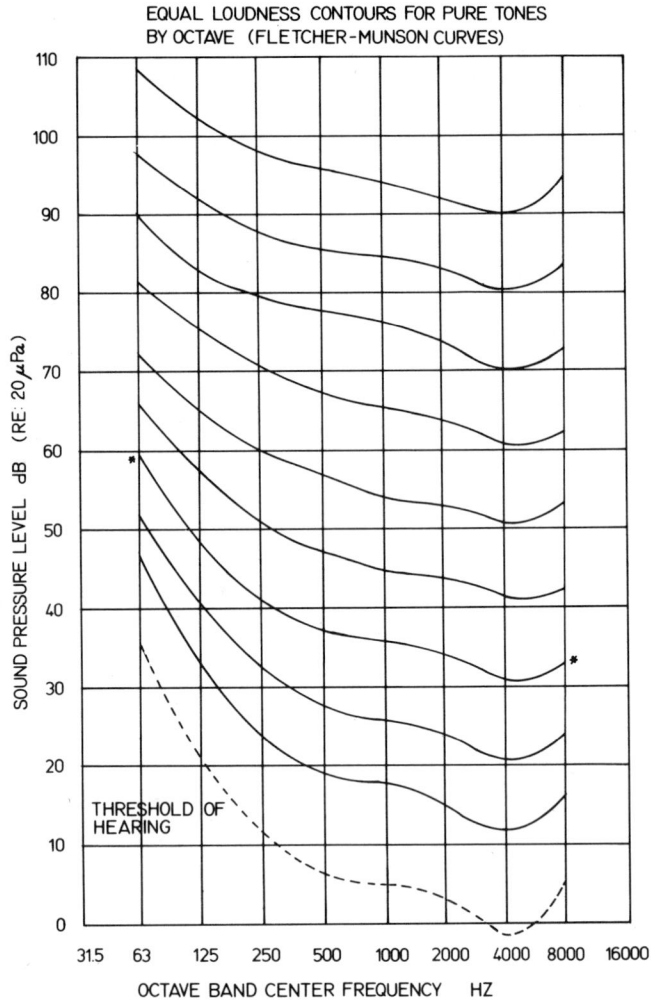

Figure 7 The Fletcher-Munson curves form the basis for many other curves and standards used in acoustics and noise control.

Will you ever need to use the Fletcher-Munson curves? Very probably, no—that is, not directly. If you continue to work in noise control you will use them, in derived forms, every day.

A-WEIGHT

For example, OSHA wants you to measure noise with an A-weighted meter. What does that mean? Well, an ideal sound meter would respond equally to at least all the frequencies that people can hear. An A-weighted meter responds to a wide range of frequencies, too—but not equally to all frequencies. In fact, it behaves much like the typical human hearing system when the noise is about as loud as quiet conversation.

Of course, the noise being measured might be quite loud. Still, the meter responds to the energy at each frequency much as our hearing does for the curve marked with stars in Figure 7. Compare that curve with the A-weight curve of Figure 8. In both cases it takes almost 30 dB more noise at 63 Hz to evoke the same response as that at 4000 Hz.

By turning a switch on the sound level meter to "A-weight," you can insert a set of resistors and capacitors into the circuit to produce this response. Origi-

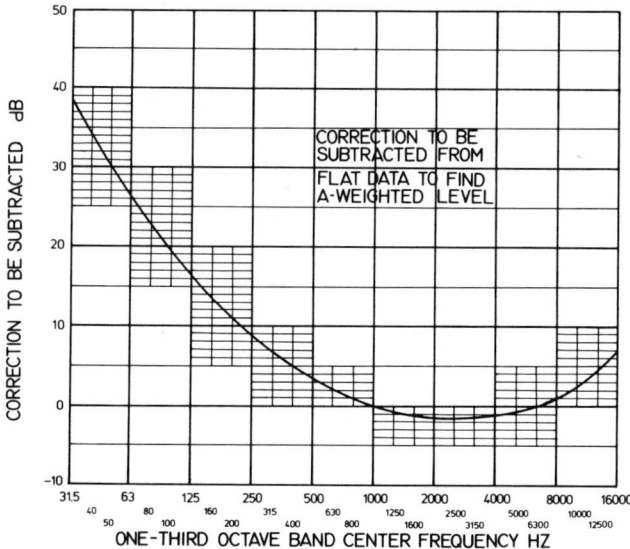

Figure 8 If the correction to be *subtracted* from the noise at any frequency in order to find A-weighted noise level is plotted against frequency it can be seen to mimic the starred curve in the Fletcher-Munson family (see Figure 7). For convenience the corrections are tabulated.

Frequency (Hz)	Correction (dB)	Frequency (Hz)	Correction (dB)	Frequency (Hz)	Correction (dB)
31.5	−39	250	−9	2,000	+1
40	−35	315	−7	2,500	+1
50	−30	400	−5	3,150	+1
63	−26	500	−3	4,000	+1
80	−23	630	−2	5,000	+1
100	−19	800	−1	6,300	0
125	−16	1,000	0	8,000	−1
160	−13	1,250	+1	10,000	−3
200	−11	1,600	+1	12,500	−4

nally, this A-weighting filter was intended to mimic the ear's ability to perceive noise when hearing conditions were good.

We continue to use A-weighted measurements today because they are thought to show a reasonable correlation to the hearing loss that may be caused by long exposure to intense noise. Even more important, a great amount of work has been based on A-weighted measurements. There is no other body of data this big on which we can draw.

There are B-weight filters and C-weight (comparable to the ear's response in noisier situations), and these days there are D-weight and N and ISO, and probably more to come.

But A-weight and the others aren't the end of the usefulness of Fletcher-Munson curves. They show up again in a family of Noise Criterion curves. The original set of NC curves has been "improved"—though some people doubt that

1971 PREFERRED NOISE CRITERIA PNC CURVES

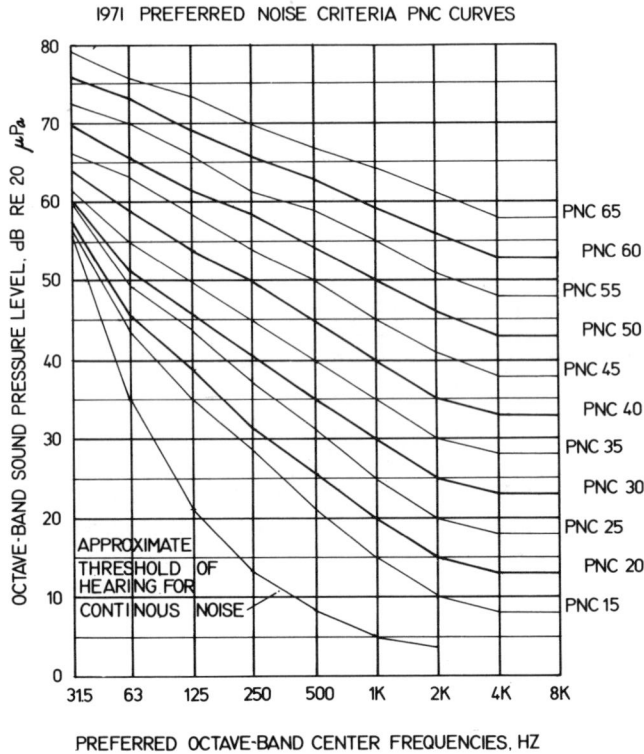

Figure 9 The preferred noise criteria curves are also similar in shape to the Fletcher-Munson family. Though not often used in industrial noise control they serve two purposes. First, they show how the sound power must be distributed in frequency if the ear is to perceive the noise as characterless. Second, they are a convenient way to describe the amount of noise tolerable in specific environments. PNC-50 is acceptable in the lobby, for example, but a large conference room ought not exceed PNC-35.

the improved curves are better than the old ones—and appear today as the PNC curves of Figure 9. The P stands for *preferred*.

These curves tell you how the ear likes to have the noise energy distributed in frequency in the noninformative background noise. They also are convenient benchmarks of how much noise is acceptable. Your bedroom, for example, ought not exceed PNC 30 but PNC 45 is perfectly acceptable in the drafting room.

When the distribution of energy with frequency follows the NC or PNC curves, the noise is relatively easy to ignore. Of course, for high PNC curves you can tell it is noisy. It may be difficult to make yourself understood or to use the telephone, for example, but the noise has little character.

If, Heaven forbid, you have to design a conference room next to the punch presses, try to arrange your noise control measures so that the background noise level in the conference room is something like one of the NC curves—the lower, the better, of course. If you don't—plus or minus 3 dB—you will get complaints. A noisier room with an NC-like background will be preferred to your quieter one, if yours doesn't have an NC-like shape.

3

DECIBELS

We've been bandying decibels about for a while without defining them. A good place to start is with those 10,000,000,000,000,000 watches. How would you like to write numbers like that (or punch them into your calculator) to solve noise control problems? Of course you wouldn't waste all that time. You'd say 10^{16}, a clear piece of shorthand. Decibels are simply a further piece of shorthand.

Decibels came about because of a problem in finding the loss of signal in a telegraph cable lo these many years ago. They weren't invented to plague people assigned to industrial noise control. We borrowed them because they are useful. If you're unhappy with decibels, relax.

If you are only going to do a little work in noise control, learn what decibels mean and master the bookkeeping drill. You will be able to get by with paper and pencil, if you have to, or with a simple four banger calculator. If you can foresee a lot of work in noise control ahead of you, bite the bullet now and buy a calculator that has the functions log and x^y (or a^n, or any other way of doing exponentiation quickly and simply). Perhaps your company will even make the investment for you.

To learn what is going on with decibels, *if you do own the fancy calculator*, do not use it for a bit but work your way through the paper and pencil drill of decibel logic.

TWO DEFINITIONS

A decibel (dB) is a number on a logarithmic scale used to express, as a level difference, the ratio of two like quantities proportional to power or energy. The ratio is expressed in decibels by multiplying its common logarithm by 10.

Don't let this thoroughgoing and official definition overwhelm you. In fact, it is far more useful than the definition that follows because you will find it tells you how to make calculations seldom called for or described.

If you are intimidated by the offical version, it might be more comforting to think of it in this way:

A change of 1 dB (upwards) means an increase in sound power of 26%. A change of 1 dB (downwards) means a decrease in sound power of 20%.

Now make a paper and pencil calculation of what this means. Start with some arbitrary power of one unit and say that the decibel level associated with that power is 0 dB. Multiply your one unit of power by 1.26 and say that the decibels have increased by 1 dB to 1 dB. Multiply 1.26 by 1.26 and see that for a power of 1.6 units, the decibels ought to be 2 and so forth. You will produce the following table:

Power	dB	Power	dB	
1	0	5	7	For each next decibel
1.26	1	6.3	8	the power must be 1.26
1.6	2	8	9	times as great.
2	3	10	10	
2.5	4	12.6	11	
3.15	5	16	12	
4	6	20	13	

There is no point in going on because the table would have *powers* identical except 10 times greater in each decade and *decibels* identical except *plus 10* in each decade.

Whether you own a calculator or not you might profit from the first surprise result of this calculation: This is a highly useful log table that you can carry in your head. If you don't think you memorize well you can always generate it by multiplying by 1.26, or—after the first three entries—by just doubling your earlier results for power ("number" in the log table, Table 2).

TABLE 2   *The Thumbnail Log Table of Acoustics*

Number (or "units of power")	Log	Decibels $\left(dB = 10 \log \dfrac{power}{reference\ power} \right)$ (referenced to 1 "unit of power")
1	0	0
1.26	0.1	1
1.6	0.2	2
2	0.3	3
2.5	0.4	4
3.15	0.5	5
4	0.6	6
5	0.7	7
6.3	0.8	8
8	0.9	9
10	1.0	10
(12.5)	(1.1)	(11)

Two useful (and unexpected) results? Yes, the second one is that the same series of ⅓ octave band frequencies is the set used in acoustics. (We'll be using them soon in estimating performances of walls).

Preferred ⅓ Octave Band	... and ...	Octave Band Frequencies
100		
125		125
160		
200		
250		250
315		
400		
500		500
630		
800		
1000		1000
1250		
1600		
2000		2000
2500		
3150		
4000		4000
5000		

These benchmark frequencies and their applications appear in a chart elsewhere in this book (Figure 5).

USING DECIBELS

$$2 + 2 = 5 \text{ (in dB)}$$
$$\text{and} \quad 0 + 0 = 3 \text{ (in dB)}$$
$$\text{but } 80 - 60 = 80 \text{ (in dB)}$$

As every schoolboy (?) knows. Well, every acoustician knows, anyway. The very simple explanation is that we do not ordinarily add and subtract decibels. *We add and subtract the powers they represent.* [There are exceptions: NR (noise reduction), IL (insertion loss), and TL (transmission loss) are measured in decibels but the decibels of these happy quantities can be added and subtracted directly, as explained in the glossary at the end of the book.]

Suppose we had a small fan in a room and that it was the only noise source. Suppose also that it produced 60 dB. If more air circulation were needed and a second, identical fan was added, what would we expect to happen to the noise? It is fairly evident that something doubled. Air flow, we hope, has doubled. The electric power consumed has doubled and so has the power being radiated by the now doubled source. But *when power doubles*—check this in the thumbnail log table—the *decibels increase by 3.* For two fans, then, 63 dB. This is why you can also write $2 + 2 = 5$, if you are working in decibels.

Suppose you ran 100 of those fans in the room. You could extend the short log table and find that for *100 times the power, the decibels increase by 20.* This is clearly shown by the first definition of decibels given earlier:

1 The ratio mentioned in the definition is (in this case) the power of 100 fans/ the power of one fan (or 100).
2 The log of that ratio is 2.
3 Ten times that log or—by definition—decibels, is 20.

For 100 fans then, 80 dB (60 dB + 20 dB = 80 db).

It becomes a little harder to work out the third "gee whiz" example of 80 dB − 60 dB = 80 dB. Let's continue to work with these same fans to do it. One fan produced 60 dB. One hundred fans produced 20 more or 80 dB. Now suppose you turned off one of the fans. What will happen to the decibels now? You could check this out with our short log table. For 99 units of power, what is the nearest decibel reading? It is still 20. Would you like a better log table? (The log of 99 is 1.9956352.) Based on that, 99 fans should have 19.95. . . decibels more than one fan. This is still 20. So turning off one fan of the hundred doesn't change the decibels. For 99 fans, then, also 80 decibels (80 dB − 60 dB = 80 dB).

OTHER METHODS

So far we have used the short log table and it is adequate for most all work with decibels. It is a little pesky, though, keeping track of the decibels associated with some arbitrary "1 unit of power" as we did with the 60 dB fan. (In the real world, of course, the fan would have produced 67 dB or some other unhandy number.)

If you have the right calculator or if you are willing to work with the big numbers, you can make use of a log table where one unit of power is associated with 0 dB. In this case the small fan has a power of 1 million; two fans, 2 million; and 100 fans, 100 million. This is very handy with the calculator.

Using a suitable calculator you'll find decibel arithmetic a snap. Two relationships, opposite sides of the same coin are used:

$$L = 10 \log W \qquad \text{and} \qquad W = 10^{L/10}$$

where L = the level in decibels
W = the power represented by the level

This is the easiest method of handling decibel arithmetic. Moreover, calculators don't get weary when there are a lot of these calculations to be done.

A good calculator with the two needed keys is just as essential as a good sound level meter in noise control work. If you (or your company) are trading up in calculator quality, a programmable calculator also will pay its way if the amount of work to be done is great.

If your calculator has LOG and 10^X (or similiar) keys you should run over the last few fan problems using the calculator.

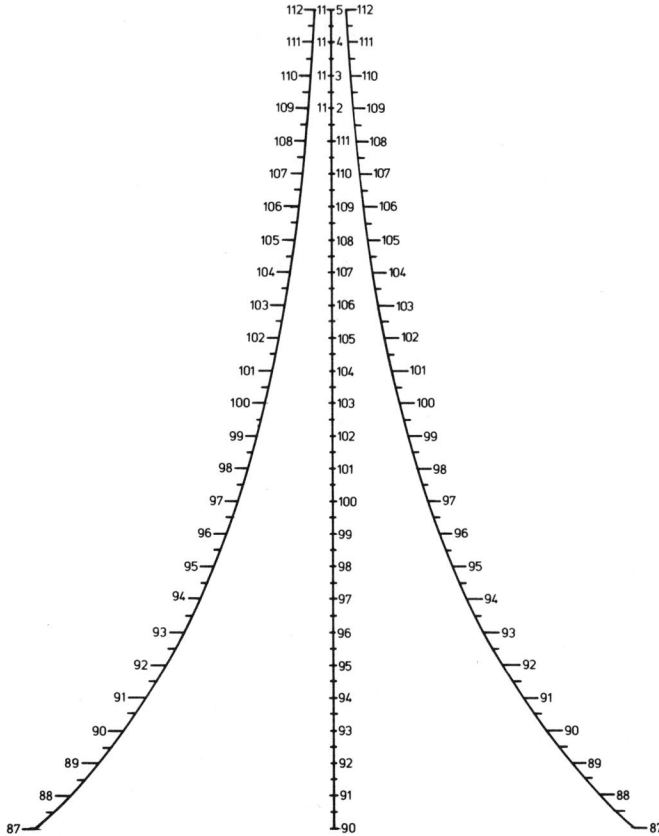

Figure 10 The decibel addition nomograph may be used to add and subtract sound pressure levels in decibels. To add two levels within its range lay a straightedge across it so that one of the levels is on one outer scale and the other contributing level is on the other outer scale. The combined level is intercepted on the center scale. Note that when the two levels differ by more than 10 dB your straightedge will cut through the curve just above the lower level. This should remind you that you need not add anything to the higher level. For repeated addition simply transfer the intermediate answer to one of the outer scales. To subtract one level from another use the center and one outer scale as input and read the answer on the other outer scale. For example, if a fan and a printing press running at the same time produce 97 dBA and the fan alone produces 91 dBA, aligning 91 on the left scale with 97 on the center scale will bring your straightedge to 96 on the right scale—showing that the press alone has a level of 96 dBA. The chart can be used for other ranges if the same constant is added to all scales.

Another method which can be used is the nomograph in Figure 10. And still another method is the use of a *difference table.* (Table 3)

TABLE 3 *Difference Table for Adding dB*

Difference between Two Levels (dB)	Add to the Higher Level (dB)
0	3
1	2.5
2 or 3	2
4	1.5
5 to 7	1
8 or 9	0.5
10 or more	0

There are two ways to use the difference table. One is often faster. The other is sometimes more accurate. Using the same data, the following worked examples are provided.

Add the following list of decibels the accurate way (starting with the smallest):

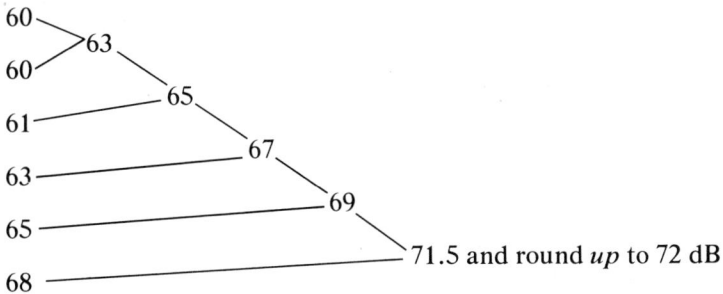

Add the same list the fast way, starting with the largest:

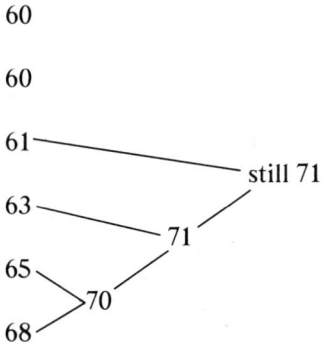

Solving with a calculator, you'll find the sum to be 71.7 dB, which should be reported as 72.

IT GETS A LITTLE WORSE

Now that we have the mechanism of decibels in hand, let's take a closer look at what we are talking about. We have been using expressions such as "the decibels increase" or "the new decibels are," which is pretty loose use of the language of acoustics. What we should say is that "the level increases" or " the new level is."

The definition of a decibel talks about a ratio of power or powerlike quantities. The basic sort of level that describes sound is a *power level L_W* (or, in the older literature PWL). It is defined as

$$L_W = 10 \log(W/W_0) \text{ (in dB)}$$

where L_W = the power level re: 10^{-12} watts* (dB)
    W = the actual power emitted by the source
    W_0 = the reference power of 10^{-12} watts*

Pay close attention here. Power level describes how much noise the source makes—how much power it radiates—but you cannot measure power level in any simple way. *It is not what you read on a sound level meter.* All the examples we worked with the small fans were based on power level. The answers we got were correct in power level and only approximated what would be read on a sound level meter. In a moment we'll see why. What does a sound level meter read, then? A quantity called *sound pressure level L_p* (or, in older books, SPL). It is also measured in decibels and hence is based on the ratio of two powerlike quantities. L_p, however, describes *how much sound power is available from the air at the location where the measurement is made.*

Power in the air is extremely hard to measure. What can be measured easily is pressure in the air. Let's explore how a "powerlike" quantity can come out of pressure (always remember we are not talking about the steady barometric pressure but rather the tiny pressure variations caused by sound waves).

From elementary physics you recall that power is a rate of doing work. Power equals work per unit time.

Work was defined in one elementary way as force acting through a distance. Work equals force times distance.

Can pressure be related to force? You bet! As soon as you define some area like the area of your eardrum or the area of a microphone diaphragm for any one pressure, there is just one force.

Can pressure be related to distance? Think now. The distance we need to consider is how far the eardrum moves or how far the microphone diaphragm moves. Both of these are proportional to pressure, too. The greater the pressure

*Be careful! Some older literature uses 10^{-13} as a reference.

variations, the farther the eardrum or diaphragm will travel. Thus the work will be related to pressure times pressure or pressure squared. And power must be related the same way.

Having established that power varies as the square of pressure, we could write

$$L_p = 10 \log(p^2/p_0{}^2) \text{ (in dB)}$$

where L_p = the sound pressure level (dB) referenced to 20 μPa
p = an average pressure variation (μPa)
p_0 = the reference pressure variation of 20 μPa

Since $(p^2/p_0{}^2)$ is the same as $(p/p_0)^2$, and since taking the log of any quantity with an exponent can be done by taking the log of the quantity and multiplying that log by the exponent, the same relation is often expressed as

$$L_p = 20 \log(p/p_0)$$

where the same definitions hold.

LET'S LEVEL

Both these quantities (L_W and L_p) are called levels and are measured in decibels. The more important one for you, as you get started in acoustics, is the sound pressure level L_p. L_p *is what you can read with a meter*, as was pointed out earlier. L_W can also be useful too, and you can *remember that it always has to be calculated.*

The last of these equations is thorny and relates the two quantities L_p and L_w:

$$L_p = L_w + 10 \log \left[\left(\frac{Q}{4\pi r^2} \right) + \left(\frac{4}{S\bar{\alpha}} \right) \right] + 10$$

where L_p and L_w are familiar
Q = the directivity factor (no units)
r = the distance between the noise source and the point of measurement (ft)
S = the surface area of the room in which the measurement is made (ft^2)
$\bar{\alpha}$ = the average absorption coefficient of the room surface (no units)

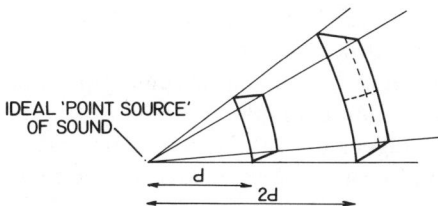

IDEAL 'POINT SOURCE'
OF SOUND

d

2d

Figure 11 The spherical spreading of sound waves for a point source makes estimating the change of level with distance easy. The level drops 6 dB every time the distance is doubled.

This book makes no direct use of this whole equation. All the effects described by variables and terms of the equation are important, however, and these are set out below.

Q

The directivity factor Q accounts for the fact that few noise sources radiate equally in all directions. In fact, it's hard to find one that does. Using only your ears you would have no trouble finding which side of a power lawn mower had the exhaust pipe, for example.

Besides the inherent directivity of the source we usually run into the directivity caused by nearby reflecting surfaces. For an ideal source hung from a sky hook in open air, Q is 1. Over a reflecting surface like a floor, it is 2. In the corner formed by two walls, Q is 4, and in the corner formed by two walls and a floor or ceiling, 8. In practice, Q may have any value—1.23 or 12—though values much greater than 10 are not common. We will be dealing with some practical considerations involving Q a little later in this book.

$4\pi r^2$

This is the equation for the surface of a sphere. Now we are getting back onto familiar ground. Remember the ever-expanding spherical bubble of sound waves. This term accounts for the way in which the sound pressure level decreases with distance.

If we had an ideal source ($Q = 1$) hung from a sky hook and used a sound level meter to measure L_p, we would find that every time we doubled the distance between the source point and the measurement point, the sound pressure level would drop 6 dB.

This holds for a "point source" of noise. Any source is a point source when the distance between the source and the measurement position is much greater than the dimensions of the source. Notice that when the distance between the source and measurement positions is doubled, the sound power is required to distribute itself over four times as much area in the expanding shell of the sound wave. The power (actually pressure squared) received by a microphone at distance d is four times as great as it is at distance $2d$. The power received by the microphone at $2d$ is called 1 unit, for convenience (Figure 11). Then

$$\Delta L_p = 10 \log(4/1) = 6 \text{ dB}$$

Often the noise source cannot be perceived as a point. Sometimes it is a line. Busy highways or noisy pipelines, for example, are "line sources," and for each doubling of distance the sound pressure level drops only 3 dB.

A common case in industrial noise control is that the measurement position (the operator's station at a large machine, for example) is so close to the machine that the machine dimensions are bigger than the distance between it and him. Here you can expect *no* reduction in sound pressure level as you move away from the source *until* the source begins to be small in comparison with the distance between your measurement point and it.

A handy way to estimate L_p at various distances from a point source is

$$\Delta L_p = 20 \log \frac{d}{d_0}$$

where ΔL_p = the change in sound pressure level (dB)
  d = the distance from the source (any units)
  d_0 = the distance for which L_p is known (same units as d)

A true line source can be evaluated by

$$\Delta L_p = 10 \log \frac{d}{d_0}$$

The effects of the $(Q/4\pi r^2)$ term of the equation have to do with the direct field of noise radiating from a source. In most of the *direct field L_p* can be expected to diminish 6 dB for a doubling of distance. Quite close to the source, the

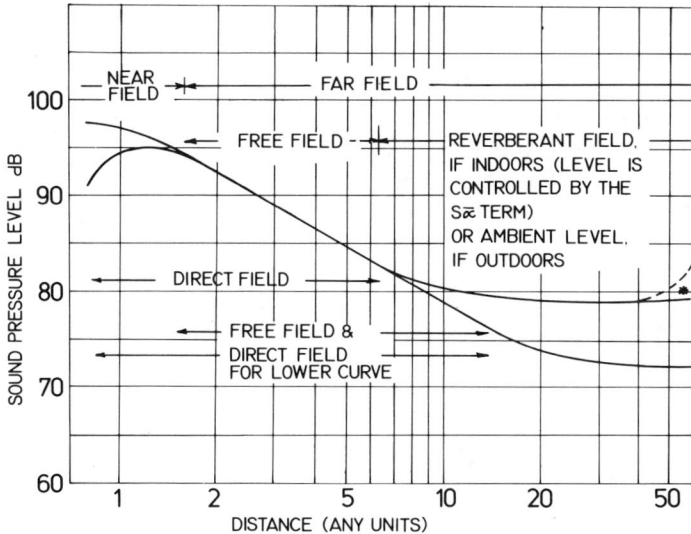

Figure 12   A generalized plot of level versus distance for a single noise source. The split in the curve at the left end may be caused by interference effects or some parts of the source may be noisier than others. The split at the right end is governed by the amount of absorption (indoors) or other noise sources like crickets or distant highways (outdoors).

6 dB effect does not hold and this region is called the *near field*. Anything beyond is the *far field*.

Sā

The generalized plot of L_p versus distance from the source takes on a new behavior at some distance when the location is indoors. This is a region where the level is fairly constant regardless of distance. It is the *reverberant field* and the level here is controlled by the environment—usually by the absorption of the room surfaces. Chapter 6 is devoted to this subject so we will touch it only lightly here.

In the *far field*, one portion of the curve is a well behaved 6 dB line and this portion is also called the *free field*. As you move farther away from the source indoors you will finally be stopped by a wall. Here odd effects may take place and the sound pressure level may actually rise because of local reflection and more complicated causes. This has been called the "far out field."

For sound sources outdoors there is a similar shape to the curve but the flattened region in this case is not due to reverberation but to the *ambient noise*. The ambient is noise that comes from all other sources—wind in the trees, crickets, distant traffic, and so on.

PROBLEMS

1 *A worked problem in space averaging.* In an effort to supply you with representative data, the manufacturer of a small gas engine has supplied you with A-weighted levels at 20 locations one meter from the surface of his engine. What A-weighted sound pressure level best describes the average level in the space 1 m from its surface?

Data

Location	L_A Level (dBA)	Location	L_A Level (dBA)
1	90	11	91
2	90	12	93
3	92	13	95
4	93	14	95
5	91	15	95
6	88	16	94
7	87	17	90
8	89	18	90
9	90	19	89
10	90	20	90

Method

(a) Regroup the data by level.

(b) Find the power corresponding to each level. The powers shown here were found with a calculator by taking 10^x, where x, in each case, was $L_A/10$. You might also use the "thumbnail log table" by making "1 unit of power" (see Table 2) the reference for 80 dBA.*

(c) Multiply the power found in (b) by the number of times it appears for each level.

(d) Sum the products found in (c).

(e) Divide by 20, the number of sample points.

(f) Convert the resulting power back to dBA.

Calculations

L_A Level (dBA)	Equivalent Power (approximate calculator answers*)	Number of Times Found	Product
87	5 $\times$ 10^8	1	5 $\times$ 10^8
88	6.3 $\times$ 10^8	1	6.3 $\times$ 10^8
89	8 $\times$ 10^8	2	16 $\times$ 10^8
90	10 $\times$ 10^8	7	70 $\times$ 10^8
91	12.5 $\times$ 10^8	2	25 $\times$ 10^8
92	16 $\times$ 10^8	1	16 $\times$ 10^8
93	20 $\times$ 10^8	2	40 $\times$ 10^8
94	25 $\times$ 10^8	1	25 $\times$ 10^8
95	32 $\times$ 10^8	3	96 $\times$ 10^8
Total samples		20	
Total power			299.3 $\times$ 10^8

(e) $\dfrac{299.3 \times 10^8}{20}$ $\dfrac{\text{(total power)}}{\text{(sample points)}} = 15 \times 10^8$ average power at each point

(f) (Using a calculator) take $10 \log(15 \times 10^8)$ and round to the nearest whole number of 92 dBA. (Using the table) The fact that you have *15* shows that the power is in the decade above the 80 dBA decade, or in the 90s. The nearest power entry to 1.5 in that decade (15 referenced to 80 dBA) is a log of 0.2, or 2 dB. The answer is therefore the same, 92 dBA.

2   Problem in space averaging. You have made measurements in a small room in six typical locations. From the following data, what is the best average level to assign to that room?

*(Using Table 2.) Your table should look like that in the calculation section except that you will have no 10^8 multipliers.

Your six level readings are: 98 dBA, 96 dBA, 92 dBA, 94 dBA, 94 dBA, and 97 dBA.

Answer

96 dBA (you found 95.6+, but report 96).

3 In steam cleaning a pump and motor that have been doused in heavy oil, it is found that a level of 108 dBA is typical at the operator's ear. If the job drags on and there is plenty of steam and enough nozzles what levels would you expect as two, three, or as many as six steam lines were brought into play? (Assume that all operators are about equally far from all nozzles.)

Answer

Number of Steam Lines	Level dBA
1	108
2	111
3	113
4	114
5	115
6	116

4 *A worked problem in time averaging.* A small reverberant room contains two pumps, an air compressor, and a fan. All operate independently and go on or off as required. Individually, these units produce the following levels in the room:

Pump 1	92 dBA	Compressor	97 dBA
Pump 2	95 dBA	Fan	93 dBA

People may enter the room at any time and stay for any length of time before leaving. In order to account for this room as part of their noise exposure, it is convenient to assign an average level to the room. Checking the history of a number of typical shifts you are able to assign typical sequences of operation as shown below. What average level will you assign this room?

Data

A typical shift consists of:

HOUR OF TYPICAL SHIFT

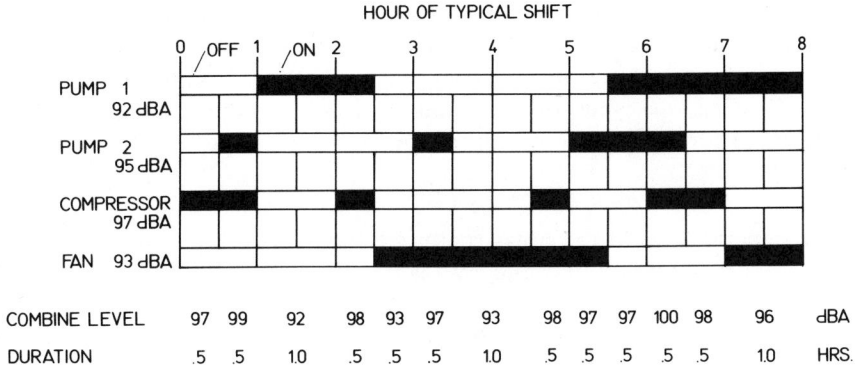

		0	OFF	1	ON	2		3		4		5		6		7		8	

PUMP 1 92 dBA

PUMP 2 95 dBA

COMPRESSOR 97 dBA

FAN 93 dBA

COMBINE LEVEL	97	99	92	98	93	97	93	98	97	97	100	98	96	dBA
DURATION	.5	.5	1.0	.5	.5	.5	1.0	.5	.5	.5	.5	.5	1.0	HRS.

Method

To find a time average for the room, multiply the power associated with each level by the time the room is at that level. Sum the time-power products and divide by the 8 hr total shift. Convert the power to a level.

Calculation

(Intermediate answers are approximations of what you will find using a calculator. If you are using the "thumbnail log table," see the notes in the first problem.)

Level dBA	Equivalent Power	Time at Level Hours	Time-Power Product
92	1.6×10^9	1.0	1.6×10^9
93	2.0×10^9	1.5	3×10^9
94	—	0	—
95	—	0	—
96	4×10^9	1.0	4×10^9
97	5×10^9	2.0	10×10^9
98	6.3×10^9	1.0	6.3×10^9
99	8×10^9	1.0	8×10^9
100	10×10^9	0.5	5×10^9

Total hours 8.0

Total time-power product 37.9×10^9

In any typical hour, therefore, you could expect a power of

$$\frac{37.9 \times 10^9}{8} \quad \text{or} \quad 4.74 \times 10^9$$

The equivalent level is 97 dBA.

Suggestions (1) Rework the problem on the assumption that any of these pieces of equipment runs continuously—or never runs. (2) Rework the problem but assign combined levels to the various equipment combinations with a precision of 0.1 dBA. Does your answer change? (3) Rework the problem finding the time average for each piece of equipment, *then* combine. This is faster and equally accurate.

5 The general level in the press room is 87 dBA with only press 1 running. When only press 2 is running, the level is 93 dBA. When all three are running, the level is 97 dBA.

   What is the level with only press 3 running? (Hint: the difference table or nomograph provide quick solutions. Do it with calculator or pencil and paper to verify the answer.)

Answer

94 dBA

   Assume that the average situation is that any two presses are usually running. What is a good level to assign to the general press room area?

Answer

95 dBA

6 *Worked problem in adding and subtracting decibels.* Two medium sized boilers are identical except that they are mirror images of each other. The space between them is at 100 dBA. Both your ears and your meter tell you that in that space the dominant noise sources are the two forced draft fans. There are many other sources in the area surrounding the boilers. The fans can be equipped with intake silencers that will reduce fan noise by 12 dBA. Will this be enough to get the area between the boilers to 97 dBA? Suppose your target is 90 dBA?

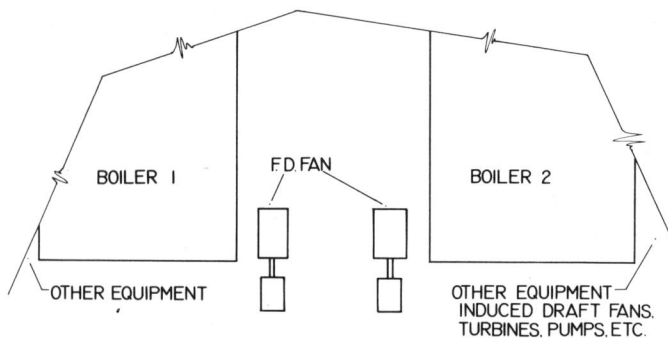

Data

You note that when one of the boilers is down the space between them is at 98 dBA. Assume that this change in noise level is caused by shutting down the forced draft fan.

Method

(a) From the two known noise levels of 100 and 98 dBA, calculate what the level would have been in the space between the boilers if the now idle forced draft fan had been the only noise source.

(b) Assume that the two fans are identical noise sources. (Are they? In the real situation you will have the chance to check this closely when they are both running.) Calculate what the noise level in the space would have been if both fans were off but all the other equipment was running normally.

(c) The background level you have just found can be recombined with the noise from the two fans *after* they have been fitted with intake silencers.

General note In a problem like this you will find that you have to break the general rule and work with fractions of decibels. In subtracting levels that are close to each other, quite small changes in the higher level result in quite large changes in the final level. There will be considerable uncertainty in your predicted background level and you must find a way to deal with it before you recommend that a substantial amount of money be spent.

Calculation

Calculation results for two methods follow. The calculator method is shown on the left; the nomograph method is shown on the right. Despite the apparent simplicity of the nomograph method, you should follow the logic of the calculator method to get the best understanding of the problem.

Step 1   (Figure 10)

Power for 100 dBA $= \quad 10 \times 10^9$ Aligning 100 on the center scale
Power for 98 dBA $= \quad 6.3 \times 10^9$ with 98 on one outer scale pro-
$\qquad$ Difference $\qquad 3.7 \times 10^9$ duces ~ 95.7 dBA on the other outer scale.

We attribute this difference to the forced draft fan, which is now shut down. If you are using the calculator, you will find 95.7 dBA by taking 10 log (difference).

Step 2

If both the forced draft fans were running and nothing else was, the power would be

$$2 \text{ (fans)} \times 3.7 \times 10^9 \text{ (per fan)}$$
$$= 7.4 \times 10^9$$

To find the background level subtract the power for the two fans from the original 100 dBA.

Power for 100 dBA $= 10 \quad \times 10^9$

Power for two fans $= \underline{\quad 7.4 \times 10^9}$

Difference $\qquad 2.6 \times 10^9$

Two fans will have a level 3 dB above the level for one fan. This will be 95.7 + 3 or 98.7. Align 98.7 on an outer scale with 100 on the center scale and read ~94.2 on the other outer scale. Since this is 100 minus the contribution of the two fans, it must be the background level.

The difference of approximately 2.6×10^9 can be converted to 94.2 dBA by multiplying its log by 10. Since it represents the noise after the two fans' contribution has been removed, it must be the background noise from other equipment. This, if true, answers the second part of the original problem. Silencing the fan intakes cannot possibly achieve 90 dBA in the space between the boilers since it will be at 94 dBA without the fans.

Step 3

If both fans were equipped with intake silencers to reduce their noise by 12 dBA,

Power from background (see step 2) $= 2.6 \times 10^9$
The new fan level is 95.7 − 12 or 83.7 and power from one fan is $10^{8.37}$
Then

Subtract 12 from the level of 98.7 for the two fans. Align the new level (86.7) on one outer scale with 94.2 on the other outer scale and read 95 dBA on the center scale.

$$\qquad\qquad 2.6 \quad \times 10^9$$
plus $2(10^{8.37}) \qquad = \underline{0.47 \times 10^9}$
New total power $\quad = \quad 3.1 \quad \times 10^9$
which converts to 95 dBA.

Thus the 97 dBA requirement can be met by silencing the fan intakes.

If the inherent error in making the measurements was ±0.5 dBA, a range of results opens up. They are summarized in the following table, which shows the extreme cases. All the figures are in dBA.

Original Levels For		Level For			Total
Both Boilers	One Boiler	One Fan	Two Fans	Background	(Fans silenced)
100.5	97.5	97.5	100.5	0?	89 (88.5)
99.5	98.5	92.5	95.5	97	97

In using this technique—especially if you explore the extremes caused by measurement error—you may find impossible results. For example, noise from the two fans might look like it exceeded the total noise. The method is worthwhile as an exploratory tool and as a way of estimating whether some noise control measures are worthwhile or quite risky investments. Methods involving subtraction of levels that are close together are the only methods where using fractions of decibels is justified.

4

INSTRUMENTS

It will be obvious—if it is not already—that this topic could easily fill a book much longer than this one. All we can hope to do here is introduce types of instruments and what they can and cannot do for you.

ONE INSTRUMENT YOU SHOULD BUY

If you could only own one acoustical instrument, a good choice would be a mechanic's stethoscope. The best ones cost about $10. Although they cannot give you any numerical data, they can give you an understanding of what is causing the noise, and often that understanding will let you find a clever (cheap!) way of solving the problem.

EXAMPLE A stethoscope with a mechanical pickup will tell you something interesting about the noise of granular material being conveyed by air through a pipe. Virtually all the noise is caused by impact at or just downstream of the elbows. Perfectly reasonable when you think about it, but it may not occur to you until you actually hear it. Result? Cover only about 20% of the pipe (at the elbows) instead of the whole run.

EXAMPLE Too much plant noise is getting into the adjacent office. Use the mechanical pickup to compare the noise radiating from the window, door, and ventilating duct (or any other surface). Compare these to the wall. Thus you may find the weak link. Then, remove the mechanical pickup and use the open tube to check leaks around the door, at the baseboard molding, and at the pipe and ductwork that penetrates the wall. You will probably be surprised at how well you can locate leaks.

One caution, however, with this or any acoustical instrument: You will always find a noise build-up in corners, due partly to Q—especially if you test the space about an inch from the surface. Allow for that build-up.

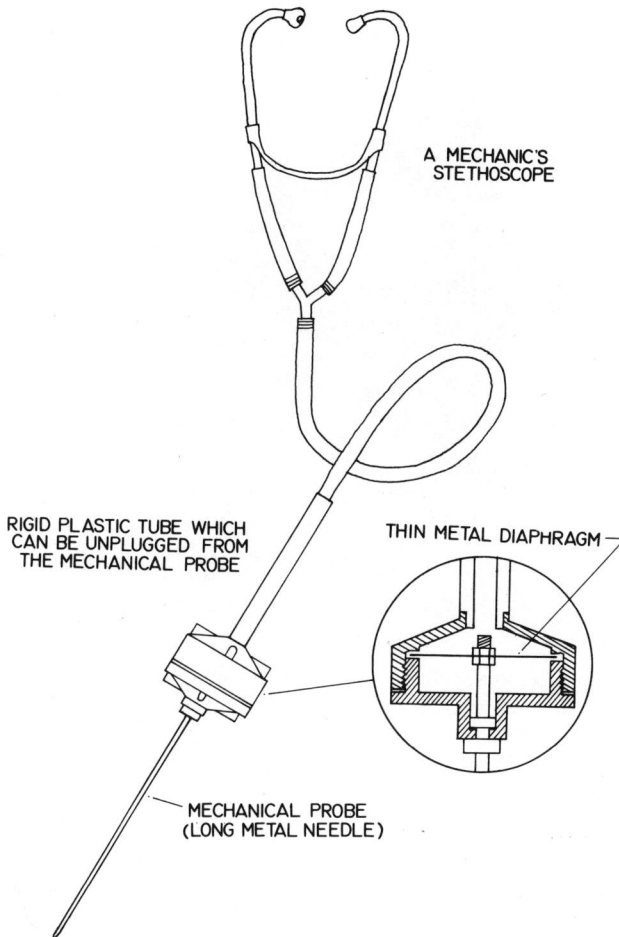

A MECHANIC'S
STETHOSCOPE

RIGID PLASTIC TUBE WHICH
CAN BE UNPLUGGED FROM
THE MECHANICAL PROBE

THIN METAL DIAPHRAGM

MECHANICAL PROBE
(LONG METAL NEEDLE)

Figure 13 The acoustical instrument you cannot afford to be without is the mechanic's stethoscope. You note there is no place on it to take a reading in numbers. The data go directly to the brain.

SURVEY METERS

What most people think you should have, if you could only have one acoustical instrument, is a survey meter. Like a lightmeter, it gives you a single number value for the sound pressure level. Usually it permits you to measure the level A-weighted and C-weighted (or "flat" or "all pass" or "linear").

Flat or *all pass* or *linear* means equal response to noise of all frequencies. C-weight is almost flat. A C-weighted meter is flat from about 100 to 4000 Hz. It falls off about 6 dB at 20 Hz and 4 dB at 10,000 Hz.

Figure 14 A popular survey meter. (It meets better specifications than "Type III, survey" described in Chapter 5. It is actually a Type II.)

Usually, also, a survey meter permits you to read with a "fast" or "slow" meter. OSHA requires you to control noise as measured A-weighted with a "slow" meter. This is a good "average" kind of measurement for steady noise. It is sometimes frustrating to come up with a single number because the noise varies or is intermittent. The usual techniques for this problem come up a little later.

Compare the A- and C-weighting curves in Figure 15. For diagnostic work with a survey meter it is a good idea to measure in both weightings. If they are the same (or if the A-weight is 1 dB higher), then the bulk of the noise energy is in the middle and high frequency end of the spectrum. If the C-weighted reading is noticeably higher than the A, the bulk of the energy must be in the lower part of the spectrum. For example, for a C-weight of 100 when the A-weight is only 80 dB, one explanation would be that the noise is all at about 100 Hz.

Before you spend your company's money for an instrument, you simply must reread this and the next chapter as a bare minimum. Talk to people who know something about the subject. Talk to instrument salesmen. Worry a little. You will be making a substantial investment and foreordaining which instruments you may later acquire.

Everyone who works at acoustics has pets and prejudices. You must read this book with the understanding that the author does too. You may discount

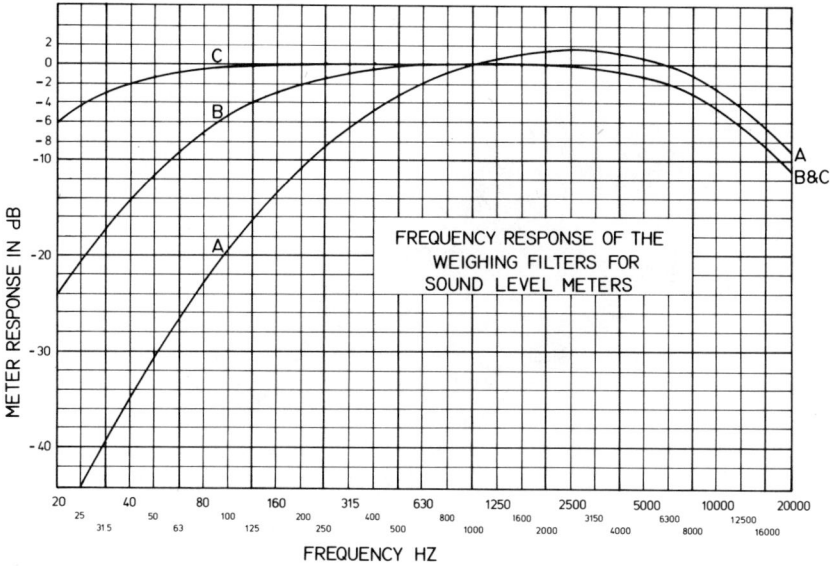

Figure 15 If you must attempt to make a diagnosis with nothing better than a survey meter take note of both A- and C-weightings. The difference between the two may give you a clue about the shape of the spectrum.

the advice you find here. It is strongly suggested that you pay attention to the next paragraph, however, and any that follow noted as "important."

Important Be very wary of bargain price survey meters! Perhaps there are some good cheap ones on the market. As this is being written instruments and solid state electronics are thundering through a millrace so that a good, inexpensive instrument is a distinct possibility for the future. Most cheap sound meters will not stand up to careful evaluation. The A-weighted filter often bears no relation to the published curve—or anything else, for that matter. Some of the cheap meters are well off the mark in sensitivity and some cannot even be adjusted. What sort of bargain will you have gotten for your company (did you save a few hundred dollars?) when the data you report foul up a job where tens of thousands of dollars—not hundreds—are at stake?

OCTAVE BAND METERS

This is the least expensive instrument that can be used for diagnostic work. It is a blunt knife, to be sure, but it is helpful in determining what is causing noise and what can be done about it. In addition to providing A- and C-weighted (or flat) overall sound pressure levels, the octave band analyzer lets you look at the sound energy in each of the octave bands—usually starting at the octave centered at 31.5 Hz and up to 8000 or 16,000 Hz.

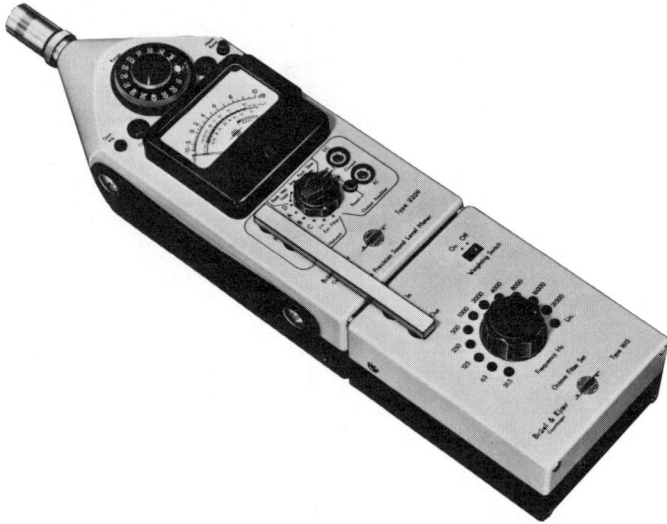

Figure 16 Two octave band analyzers. In the meter at the top the octave filter set (bottom section) is separable. The octave filters are built right into the meter at the left.

For example, where a noisy motor (mostly low and middle frequencies) is driving a conveyor belt which produces a pronounced squeal from a misaligned bearing or something, the octave band meter at successive settings of 2000, 4000, and 8000 Hz will give you a level for the bearing. The lower bands will tell you mostly about the motor. You can mathematically calculate what will happen to the A-weighted level if either of these sources is quieted. You cannot do that with a survey meter.

CONSTANT PERCENTAGE BANDWIDTH ANALYZERS

Some measurements must be made in bands narrower than one octave. Measurement of the effect of a wall (transmission loss) is always made in one-third octave bands, for example. One-third octave band measurements are also better for diagnostic work—the instrument is a sharp knife. It often permits you to pin down one pair of gears, for example, or decide whether the bandsaw blade or the fan in the dust collector attached to the saw is the real culprit.

Sometimes an instrument of this kind is tunable—infinitely variable tuning in frequency. Often it can be set for bandwidths of one-third *or* one-tenth octave. At a setting of one-tenth octave, it becomes a razor blade and you rarely will need better resolution in any real acoustic problem. However, all this resolution comes at a price!

Quite aside from the price tag, the cost in time-consuming measurements is discouraging. Slowly and carefully you tune away and jot down data, and somehow it seems that these measurements are always required someplace where it's hot and dusty, or where you have to balance on a ladder to make them.

CONSTANT BANDWIDTH SPECTRUM ANALYZERS

In the analogy we've been using, this one is the scalpel. It is not possible to do better in resolution than to use one of these. These analyzers usually display the noise spectrum on a cathode ray tube and permit the frequency range to be adjusted, typically from as low as 0 to 250 Hz to 0 to 25,000 Hz or more. In operation, they split the range into "lines" so that the whole range is displayed as 250 parts (more for some instruments). For a range of 0 to 250 Hz, each line would be 1 Hz wide.

On the screen of the tube, a continuous line traces out the spectrum. Provision is made for reading out the frequency of any feature of interest. On some of these instruments it is possible to indicate all the harmonics of any noise peak right on the screen. They can also average noise levels if these vary with time, and they can retain the highest peak reached at all frequencies if that best serves

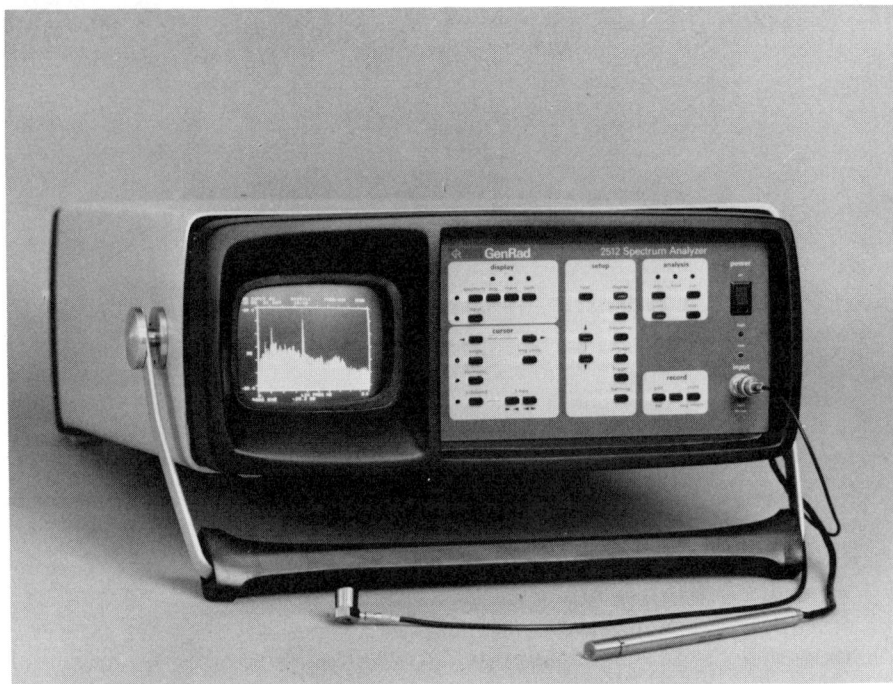

Figure 17 Spectrum analyzers can be used in the field but are more often used in the lab to make high resolution analysis of taped data.

your purpose. They usually have provision to drive an X-Y plotter for a permanent record.

One drawback of spectrum analyzers is cost. Think in terms of $10,000 when you look into this type of instrumentation. Also you will need some auxiliaries, like X-Y plotters.

Another drawback of the spectrum analyzer is weight and bulk. Progress is being made here, but even the smallest and lightest is no dream to use in the field. Although all the instruments previously mentioned are battery powered or at least can be battery powered, the spectral analyzer almost always requires line power—another drawback for field use.

The normal home of this instrument is the lab. Field data are brought back on tape and played into the analyzer. Tape is advantageous because as analysis of noise progresses, you often want to look at the signal in new ways. The spectrum analyzer is great for this and frequently coaxes the acoustican into long, thoughtful sessions about what is on the screen.

REAL TIME ANALYZERS

These are exceedingly useful devices. They can read either one-third octave bands of frequency or constant increments of frequency as explained previously. In effect, they are a series of sound level meters running continuously, each tuned to a separate band. Until 1976, they belonged in the lab and not in the field. They were big and heavy and required 110 v power.

Then, about a decade after the first one was made, a little hand-held model was introduced. This model displays the running level in each of ten octaves. Although it lacks the hard-copy output that all the lab-sized analyzers provide, it is a treat to use in the field. You do need to get used to reading it, however.

Still more recently a book-sized one-third octave real time analyzer has been produced by the same manufacturer. To anyone who has worked in the field of acoustics for a long time, it is hard to believe how much this instrument can do. It will display real time octave or one-third octave spectra, linear or A-weighted, it will "freeze" and remember a spectrum (or two spectra in separate memories) even after it has been turned off, and it can be fitted with an accessory to drive an X-Y plotter. Moreover, it can be held comfortably in one hand. Ten years ago you couldn't have carried this much equipment on your back or even in a wheelbarrow.

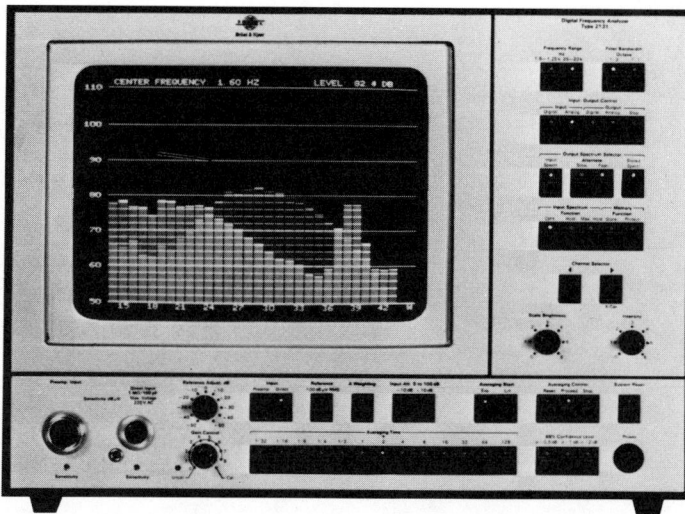

Figure 18   A real time analyzer displaying a spectrum in one-third octave bands. Many functions are available within these instruments. They can A-weight the data or average it over time or find the maximum reached in each band.

(a)

(b)

Figure 19 Two highly portable real time analyzers. The smaller displays levels in the octaves/from 31.5 to 16 kHz. The tape cassette shows scale. The larger, although easily carried in the plant, can display octave or one-third octaves, A-weight the display, or find maximum levels reached as the larger lab models do.

LEVEL RECORDERS

These are devices used with various analyzers to produce a strip chart record either A-weighted (or C or flat) level with time or to show levels in individual bands.

TAPE RECORDERS

We mentioned previously that this is a useful way to get data back to a home base where it can be analyzed carefully. Sometimes it is the only practical way to take data. Given the job of documenting levels throughout an 8-hr shift, for example, even the most conscientious technician will make mistakes in recording data. The tape recorder? You had better believe it does, too, and sometimes they're beauts.

You can't do much serious work in acoustics with a $29.95 cassette recorder. It is handy for taking notes in the field, and, if the batteries are fresh, it's good enough to let you pick out big frequency peaks within its limited range. Don't believe anything it tells you about level.

Most taping is done with precision tape recorders made for use in acoustics. One manufacturer is so prominent—probably accounting for 90% plus of all tape recorders used in acoustics—that you should look to its instruments first if you decide to go this way. The tape recorder is a Nagra, made in Switzerland by Kudelski. Some models have a sound level meter built right in.

Important On some of the older models of Nagra tape recorders with built-in sound level meters, an undesirable type of detector was used. The quality of the noise influences the reading. The output of the instrument is not affected. In other words, the built-in meter is useful only as a rough gauge of level. The proper information is *only* available by reading the output of the tape with a good meter (details follow in the section on type of detector).

These instruments, depending on what features you want, cost from about $3500 to $5000. They are light enough to carry through the plant, but they do become cumbersome before the day is over. They weigh 20 lb, but they feel more like 50 lb toward the end of the day.

Reliability is not the long suit of tape recorders (or most sound level meters). All acoustical instruments should be treated like the prima donnas they are. They should be protected from rough handling, heat, dirt, moisture, and color schemes that clash. Seriously, you will come to wonder how and why they can break down so often—which leads to a real essential of acoustical instrumentation . . . .

EARPHONES

Every good acoustical measurement device (with the exception of the real time analyzers we looked at) makes provision for driving a pair of earphones.

Golden information   Always monitor your instruments by listening to the output with earphones!*

If you are using a complex setup (meter as preamp into a tape recorder), listen to the last output. That way you can hear anything going wrong in the system.

What can go wrong? Most commonly the signal just disappears. If you have jiggled all the cable connections and tried putting in fresh batteries you will probably do what everyone else does—sweat. If your budget permits it, you will have spares of the components that most often break down (microphones and preamps). When you have tried everything else without success, a karate chop sometimes has a sanguinary effect.

However, loss of signal is not the only trouble you will pick up on the phones. Infrequently, a squeal or beeping will appear in the phones. This is fine *if* when you take the phones off you can still hear it. But sometimes it is not there in the airborne sound. The meter circuitry is usually the culprit and you should have it repaired. Feedback from your earphones sometimes causes squealing, too.

You can sometimes salvage taped data that has been written over by a beeping meter. Provided you don't have to take the data into court, provided you are not much interested in the real signals with the same frequency as the beep, and provided you have some sort of analyzer, you may be able to believe the data for all the bands except the one that carries the beep.

Distortion of the signal is common and is often a sign of low batteries. If you make measurements in a high wind (anything above 8 mph for low sound pressure levels) you will hear wind pops in the earphones. These are due to the wind shaking the diaphragm of your microphone. *A windscreen is recommended at all times* (that way you don't have to walk back to the office for it when you decide you should take a reading at the fan discharge; besides, using a windscreen has a negligible effect on the data and it helps protect the microphone, which is a most delicate and expensive piece of equipment). You can "read around" the wind pops if you are writing down the data. A tape recorder or analyzer cannot.

If you are working near arc furnaces or induction furnaces or radio transmitters, your setup is liable to pick up the power being radiated electrically rather than acoustically. This can be very confusing. In the case of the radio

---

*Not all earphones will work with all instruments, however. Phones you would typically use with your stereo set at home present a load of only 6 or 8 ohms. This means that they require a fair amount of current. Some instruments, on the other hand, are designed to work into a load of 600 ohms or a much higher impedance. If the instrument manufacturer supplies earphones with the instrument, they will be matched. If he does not or if you do not like the phones (or their cost), you can buy a transformer about the size of a sugar cube to match the output and load. Mount the transformer inside the headset so that the instrument output goes to the side of the transformer rated in thousands of ohms (2 to 10 kΩ, not critical) and the earphone speakers go to the side of the transformer rated at 3 to 10 ohms. Both your instrument suppliers and local electronics hobby shop will be helpful in getting you set up. Using low impedance phones on some meters designed to work into 600 ohms or more can result in a totally incorrect meter reading.

station, at least, your earphones will immediately explain meter readings that don't seem to match what you can hear. For other suspect electric equipment, try listening through the phones with one ear and to the airborne noise with the other—see whether they match. You still may have some doubts.

If you are using a condenser microphone (not an electret—often called a condenser mike), you will have a ubiquitous foe in making measurements. In the phones you will hear "tck—sss" and the signal will drop out only to come back and do it again and again; your foe is water. Moisture condenses in the space between the microphone diaphragm and backplate and the 200 V or so required to make the microphone work is arcing across the gap. You can break out a spare mike, rig up an electrical heater run from an automobile battery as is sometimes done, or gracefully retire. When moisture attacks a condenser mike, you might as well quit. Do not open the mike and attempt to dry it! The manufacturer can spot a fingerprint on the diaphragm at 10 paces and it voids your warranty—at a cost to you of about $500. If it is a true condenser mike and not an electret, you can bake it in an oven at 250°F for 1 hr. The moisture will have boiled off, and the microphone will be unharmed.

MICROPHONES

In selecting your instrumentation, microphones should have some consideration. More often than not, when you choose a particular instrument, you are locked into the type of microphone furnished with it—but this is not always true. Historically, there are many interesting mikes. The following discussion will include only those microphones usually used in acoustics.

Condenser

The *condenser* mike has the best reputation for fidelity and stability. Like all mikes, it has a limited frequency range. At the low end it can sometimes go all

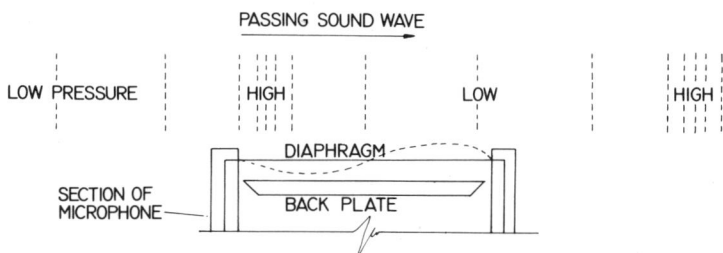

Figure 20 Microphone size limits the high frequency response. When a frequency that produces a wavelength of twice the microphone diameter is reached, the same wave is both depressing and lifting the diaphragm. Since most of the power is cancelled out, the microphone's response is much too low.

the way to 0 Hz. The high end is governed by size of the microphone diaphragm. When the wavelength of sound in the air approaches the diameter of the diaphragm, the same sound wave will be forcing one side of the diaphragm down and simultaneously lifting the other. This is common to all mikes. For high frequency (short wavelengths), use a smaller diameter microphone.

The output of a condenser mike is fairly small and its impedance is high. This means that if any length of cable is used between the mike and the measuring instrument, a preamp at the mike is required.

In addition, a condenser mike requires a high polarizing voltage. There is a relationship $Q = CE$ where Q is a charge in coulombs, C is the capacitance of a capacitor (used to be called a condenser), and E is voltage. The charge in a condenser mike is supplied by the polarizing voltage.

When the diaphragm trembles under the influence of the sound wave, the spacing between the diaphragm and the backplate changes, changing the capacitance. Since Q is a constant, the voltage between the diaphragm and the backplate must change to mimic exactly the motion of the diaphragm.

Electret

In an *electret* mike the same method is used to produce a voltage except that no polarizing voltage is used. A charge is "frozen" into a plastic material by subjecting it to a strong electrostatic field as it solidifies from the liquid to solid state. This material is aluminized and used as the diaphragm. When a sound wave makes it tremble, a mimicing voltage appears between it and the backplate.

Because there is no high voltage involved, electrets are not as likely to have fits under conditions of high humidity or condensation. On the other hand, critics of the electret mike wonder how stable it is. Does the plastic material degrade with time? Does that frozen charge leak away? Tests at the National Bureau of Standards suggest that under the worst conditions a condenser mike might drift in sensitivity by less than 1 dB in several months or a year.

The same tests show the electret drifting 2 or 3 dB in the same time. Also, while the metal foil diaphragm of a true condenser mike is hardly affected by temperature or humidity less than that which causes arcing, the plastic diaphragm does seem to be affected and will drift by a fraction of a decibel due to temperature and humidity.

The electret has an inherently high output—that is, it is very sensitive. This is not an important consideration for two reasons. First, it is also a high impedance device and requires a preamp when used on a cable. Second, much of the inherent sensitivity is lost by designing the electret for acoustic measurement to have an extended range and good linearity.

Piezoelectric

Piezoelectric mikes used to be called crystal mikes. In these, movement of the diaphragm is made to squeeze or bend or twist a material that produces a voltage when it is deformed. The material used to be rochelle salts—a water soluble chemical. Crystal mikes of this sort are quite unsuited for acoustic measurements.

In the last 15 years a much more stable material has been used. This is lead zirconate titanate, a tough insoluble material that resembles the fired clay of a flowerpot. This is a fairly rugged microphone of good sensitivity. It tolerates high humidity and condensation very well *for short periods of time.* If used in high humidity for long periods (weeks), it may change its characteristics or just stop working.

Dynamic

Dynamic microphones are the least desirable type for acoustic work. They are built much like a permanent magnet loudspeaker. In fact, you can use a PM speaker as a mike. The sound wave makes the diaphragm move. A coil attached to it and located in the field of a magnet generates a voltage as the diaphragm moves.

If the dynamic mike has anything to recommend it, it is that it has a low output impedance and can be used on a cable without a preamplifier. As a rule, neither frequency range nor linearity of dynamic mikes is very good. Though they are still used on some instruments, dynamic mikes should probably make you skeptical about the instrument that employs them.

As noted, there are other kinds of microphones, but they are not used on instruments. In general, people working in acoustics prefer microphones in the order they are presented here (i.e., condenser, electret, piezoelectric, and dynamic).

VIEWPOINTS

If a little philosophy about acoustics can be tolerated, the difference between electret and condenser mikes is the subject of mild debate among people who work at noise control. This debate illustrates a "style" of approach. The advocate of the condenser says, "It doesn't drift, it is a no-compromise design, it's the best you can get." His or her adversary says, "Yes, but it packs up at the first sign of rain or condensation, it costs you a lot of weight in the instrument to generate that high polarizing voltage, and as far as drifting is concerned that's why they make calibrators!"

The same frame of mind will lead one acoustician to lug 300 lb of equipment to the job so that he can get everything analyzed exquisitely. His counterpart may use minimum equipment and a lot of intuition. As you choose your instruments these two frames of mind deserve your cogitation. Sure, some people are lazy and some are compulsive about their jobs. Allow for that and look at some of the other factors—reliability of a simple versus a complex instrument system; the possibility of missing something important, even vital, because the instrumentation is too crude; even dodging behind a bushel of pointless data rather than coming to grips with the problem. There is no moral here, only a question.

CALIBRATING

Even if you are buying a cheap survey meter—in fact, *especially* if you are buying a cheap survey meter—you should have a good calibrator. They come in two types.

The *pistonphone* calibrator is an absolute standard. A small electric motor drives a pair of tiny ceramic pistons in and out of a chamber 250 times per second. The driving linkage ensures that the displacement of the pistons is always the same. Since the chamber is plugged by the microphone, the contained air is trapped and the pressure variation must always be the same.

The pistonphone always produces 124 dB at a frequency near 250 Hz. A new one costs about $900. If you buy or have one, don't let the airport security people open it. Exposing the mechanism damages the pistons and you will be in for long, costly repairs.

The many other calibrators are much alike. They employ an electronic oscillator (or sometimes several to produce a set of standard frequencies) which is coupled into the chamber by a small loudspeaker. These are light, handy, and somewhat less expensive than the piston type. The one that produces a series of frequencies (125, 250, 500, 1000, and 2000 Hz) is very attractive for checking any microphone or meter suspected of drifting. However, like the meter it is checking, this sort of calibrator is a circuit and may be subject to drift itself. Therefore, it needs to be checked from time to time. The output of these calibrators is usually 94 or 114 dB.

On some older meters there is a setting labeled "calibrate" and turning the switch to this setting is supposed to produce a certain reading. If any other reading is obtained, a gain adjustment can be manipulated to get the right reading.

Good idea? It only sounds good. In the first place the microphone is not included in this test. In the second, some of the instruments that use this method simply feed back some of the output of the instrument amplifier to the input. The "electrical calibration," as it is called, usually works but is not completely trustworthy. Certainly if a real calibrator and this electrical calibrator produce different results, you will know which to prefer.

5

MEASUREMENT

You have probably concluded that the first instrument you will need for solving industrial noise problems is an octave band meter. Many people do. A one-third octave meter will give you better resolution: it will also slow you up with three times as many readings to make. More philosophical questions!

When you have reached that decision, whose instrument will you buy? You needn't ignore the obvious things like styling, arrangement, and simplicity or complexity of controls and, if you are going to carry it, check the weight. What type or types of electrical connectors does it use?

There seems to be a conspiracy to require a different connector for every purpose. This can become a nuisance, and worse, as your equipment list gets larger. You will have a big box of matching cables (patches) like a male BNC going to a seven-pin DIN, or maybe a mike plug to a double banana. But it won't matter how many of these patches you've made up. You will always find you need one more.

MERIT FIGURE: TOLERANCE

There is a tolerance (or inherent error) in making any measurement. In making measurements of sound pressure level, two main sources of error are inherent. (Many more can result from sloppy technique, but here we are concerned only with the inherent, unavoidable error.) The meter itself will not have a perfectly smooth response curve with frequency. The amount by which a meter may depart from the ideal response curve determines the meter *type*. The other inherent source of error involves the direction from which the sound wave approaches the microphone. Microphone response is not uniform with direction.

The three *types* of sound level meters are*:

Type I	Precision
Type II	General purpose
Type III	Survey

*As this book is being written another wave of change is breaking in acoustical measuring and analyzing instruments. In addition to being lighter, more convenient, and cheaper than the old instruments some of the new ones appear to be much better. The classification scheme given here may be revised shortly on this account.

In making a reading of the A-weighted level of a signal with random incidence (equal sound power from every angle), the allowable tolerance for meters of each type is shown in Figure 21.

Incidence angle error has to be added to the tolerance allowed for a meter with random incidence sound impinging on it. To be honest, you will recognize the screamingly bad cases here. If all the noise comes from a point, you will either point the microphone at that point (if you are using a free-field microphone) or hold it at a 70° angle (or whatever angle your owner's manual tells you is best if you are using random incidence mike).

If there is no way to tell where the noise comes from, the random incidence idea probably holds up fairly well and you will not have to worry about the following errors. In any event, put down what you are learning in this section of this book as "general information" and what it says in the owner's manual that comes with your sound level meter as "golden information".

If you cannot find the right angle for the microphone and the worst happens to you (as when you are in the near field of large sound sources) the tolerance of the system gets worse. (See Table 4.)

Now in finding the A-weighted level someplace, add up what can happen to you in a bad, but plausible case with each of the three types of meter.

For *Type I*, assuming that the controlling frequency of the noise is around 4000 Hz, the meter will have a tolerance of $+3.5, -3$ dBA. The correct reading of 90 dBA may be shown by your *precision sound level meter* as 87 to 93 or 94 dBA.

For *Type II*, the limits are $+4, -6$ and the reading you find will fall between 84 to 94 dBA.

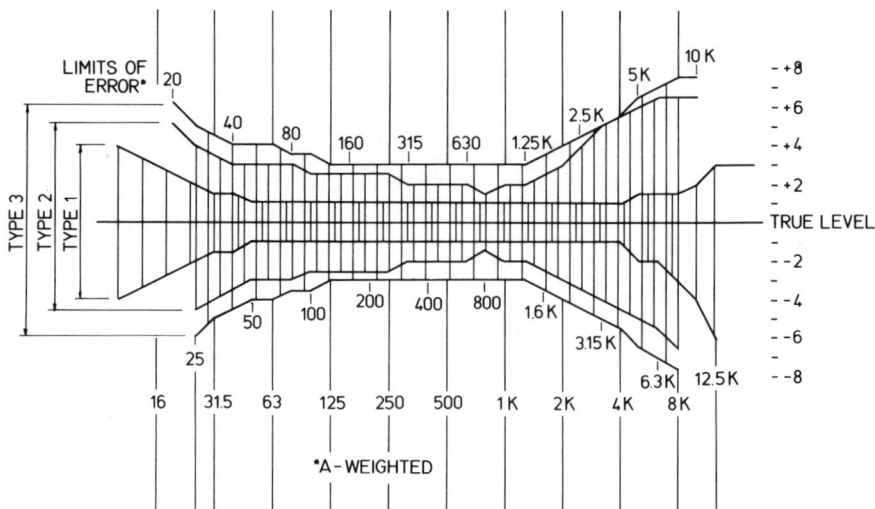

Figure 21 The allowable inherent meter error for each frequency and each meter type. These errors can be important even with meters of good quality. Poorer meters can be so misleading as to be worthless. Noise control work requires at least a Type II instrument.

Table 4 *Additional Tolerance That Results if Dominant Sound Energy Arrives at the Microphone From a Discrete Angle*

Frequency (Hz)	Type I	Type II	Type III
31.5 to 2,000	+1.5, −1	+3	±5
2,000 to 4,000	+2.5, −2	+3, −4	±8
4,000 to 5,000	+3.5, −3	+4, −6	±9
5,000 to 6,300	+4, −4	+5, −8	±12
6,300 to 8,000	+5.5, −5.5	+7, −9	±15
8,000 to 10,000	+6.5, −8		
10,000 to 12,500	+7.5, −11		

Add these figures to the tolerances shown in Figure 21.

You should avoid using a *Type III* meter if possible. If you really must use one, anything between 81 and 99 on your meter may be 90 dBA. You may get to work lowering the noise level of an 81 dBA source, or, with some other meter of the Type III ilk, you may ignore a 99 dBA source.

MERIT FIGURE

Type of Detector

If this little section is not the most obscure in this book, it is, at least, in the running. We are not going to do it justice! Sound is not a continuous pressure but a varying one. We have mentioned sine waves (which are pure tones), harmonics, and steep wave fronts (fast rise times).

To cut through a lot of fog, when you have to compare a varying quantity with a static one (the dynamic pressure of sound versus the steady barometric pressure of the air), a scheme has arisen to make the mathematics convincingly simple. It says that the steady continuous pressure equivalent to the varying pressure can be found by sampling the varying pressure at every instant, squaring the value you find, summing all the squares, dividing it by the number of samples taken, and finding the square root. It can never be done exactly. We keep pretending that it can. Some detectors, in sound level meters, do it well enough.

The consequence of not doing it well is getting the wrong answer. The meter in the old Nagra tape recorder (SJ Model) didn't do it very well and incurred an error. Cheap sound level meters also incur the error.

The cheapest sound level meter will get the right answer when the pressure fluctuations of sound are sine waves. This means that you can use any meter you like to measure the noise produced by a softly blown flute or a harp played softly.

Is that what you expect to measure at your plant? Where noise is a problem, you can expect anything but sine waves and unless your meter does a convincing imitation of running through the process outlined above to find the root mean square (rms) level, it is lying.

There are several types of detectors. The best of them is the "true" rms detector (it's almost true) and they range down through "quasi-rms," "four point," and so forth.

Consequently, to the extent that you have to measure "harsh" noise (which generally means noise with steep wavefronts and many harmonics), your meter can't tell you the truth unless it has an expensive detector, the "true rms" being the best. Often, the noise you have to measure will not be "harsh," and other meters can do it.

To return to the purpose of this section—advice on what instrumentation to purchase—get the best detector you can. Suspect instruments whose makers (or salesperson) cannot, or will not, tell how the alternating signal is converted to a steady meter reading.

Crest Factor

If you are buying a good instrument, there is a number given in the specifications and you may be able to avoid embarrassing or confusing the salesperson. Ask if there is a *crest factor* rating for the instrument. (You are covering the same ground we just examined, but didn't plow.)

The crest factor in decibels (between sine and "harsh" waveforms) considers the whole instrument—though the detector is the principal bind—and tells you how bad the sound wave can be and still give you a reading that represents the true power. The higher the crest factor, the better.

One quality instrument has a crest factor of 24. Another instrument has a much higher crest factor of 40. In nearly all cases, the values read by these two meters will be the same. Some inexpensive sound level meters have crest factors as low as 3 dB—they only read sine waves correctly.

Instruments this good are expensive, but you need good instruments in order *to diagnose* what sometimes causes noise.

Fast

Fast is way of reading through intermittent noise to investigate what a steady noise source contributes. It means that the time constant of the instrument is 200 msec. That is, if the pressure abruptly changed from one dynamic pressure to another and stayed there 200 msec before returning to its original level, the instrument reported the difference in level, assuming that its response in the meter and driving circuit were no slower than anything else in the meter.

Use "fast" to find what you can about steady noise lurking behind higher levels that were intermittent and don't worry about where all this comes from. We are getting close, here, to the difference between science and engineering.

Slow

Slow means that the meter had 500 msec to get a grip on things. In that amount of time (for steady state noise sources) it does well enough.

Impulse

Impulse noise is supposed to represent what happened in the interesting 35 msec. Well, it is a place you can set your meter, anyway. (European countries make use of this setting and write laws where it is important.)

Peak

Impact (or peak) is somewhat more understandable. The meter has a rise time of about 20 μsec and drives the peak it found into an amplifier. The amplifier, in turn, drives a "remember" circuit that tells you the peak reached. Have more faith in "peak" than in "impulse".

MAKING MEASUREMENTS

Shiny new meter in hand, we march out to the job site. Call it "snooping", don't call it measurement. You can poke around and learn some interesting things. A high reading in an unexpected part of the machine? What does it sound like in the earphones? Can you find a peak in the spectrum? Maybe it will tell you something useful about how the noise is generated. But this *is not* what we usually mean by making measurements.

A good way to start is to leave your meter in the front office and visit the job site with a pad of crossed ruled paper. Watch where people stand or go. Try to get an idea of the main noise sources (your stethoscope is probably better than your meter for this). Then make a neat sketch with the key dimensions on it and decide where you are going to make measurements.

You will certainly want to include the operator's position if you are solving an OSHA noise problem. Locations near the sources within the machine are very useful, too. If possible, catch the machine you are working on when it is shut down or running idle (without feed) but while all the normal operations of other machines continue. On your sketch, label the locations and make a table of frequencies (assuming you are using an octave band meter). *Then* begin to measure (see Figure 22).

BAD CASES OF THE JITTERS

For low frequencies, it is quite usual for the needle to roam all over the meter face. The first step, of course, is to use a "slow" meter setting but this rarely

NORTH

SPROCKET 15 HP DRIVE

↑ NEAREST WALL ~ 28' FROM CONV'R

— OVERHEAD HOPPER

~ 16'

BOX CONV'R →

SIX AUTOMATIC SCALES
& BOX FILLERS
('ACCUTROL'?)

Ⓔ   Ⓑ

Ⓒ   TURNTABLE & CONVEYOR TO
CARTON FILLER. AIR JET
AT Ⓒ HELPS MOVE BOXES

Ⓓ

Ⓑ   DESK &
Ⓐ   TEST SCALE

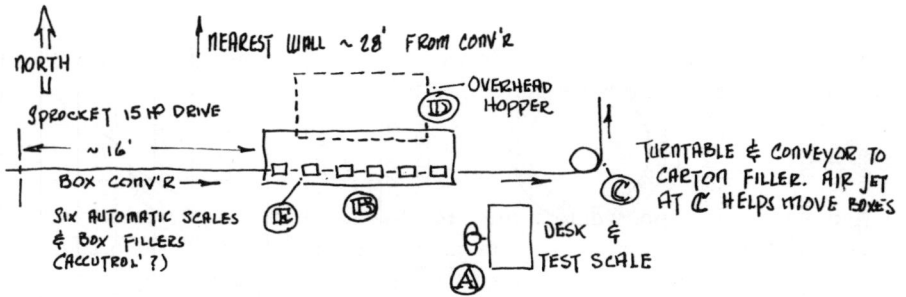

BOX FILLING OPERATION

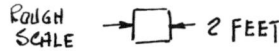

ROUGH SCALE → ▭ ⊢ 2 FEET

LOCATION	A	B	C	D	E
	USUAL OPER.	OTHER OPER.	NEAR FIELD	NEAR FIELD	NEAR FIELD OF
FREQ	Pos'n (80%	Pos'n (15%	(6") OF JET	OF BIN SHAKER	WEIGH HOPPER
	OF TIME)	OF TIME)		& HOPPER WALL	(IMPACT/DOOR SLAM)
32					
63					
125					
250					
500					
1000					
2000					
4000					
8000					
LINEAR					
A-WT					

NOTE: CEILING PAINTED CONCRETE AT 16'. NEAREST NOISY EQPT IS A
SIMILAR UNIT 35' SOUTH. FLOOR IS CONCRETE.
AMBIENT (ALL FILLERS DOWN) IS 78 dBA — MOSTLY SPACE HEATERS
WITH PEAK AT 250 HZ.

Figure 22 Take the time to make a sketch and record key dimensions *before* you start to take data. Six months later you'll be so glad you did.

subdues the jitters. There are about three basic ways to get a single value reading under these conditions.

There is the *eyeball average* which, if the excursion of the needle doesn't exceed 6 dB, means just what it says. If the needle regularly swings further than that, it will be safe to use a value 3 dB below the *usual* maximum reading.

The *best estimate of central tendency* is much used by consultants because it sounds so expensive. Basically it is the eyeball average but it includes another step. First, watch the maximum readings long enough to exclude the unusual ones. This will produce the same usual maximum we had a moment ago. Then do the same thing with the lower limit. Obviously the central tendency is what lies midway between these. (It's still safer to report 3 dB below the usual maximum if the two "usual" figures differ by more than 6 dB.)

You might note that the expression "best estimate of central tendency" is sometimes used to mean that a detailed statistical study of the data has been made. More often, the term has the casual meaning.

The *1-2-3 method* is for the timid or intimidated. Use it when the meter is really wild. Count to yourself "1-2-3" and write down the level at the instant you say "three." Do this over and over (not less than 10 times) and if you have made a set of tally marks next to your levels, you should have no trouble assigning a level to the location. It will look like this:

63 Hz Band Level, dB	
78	I
79	
80	IIII
81	⊔⊔⊔
82	⊔⊔⊔ II
83	IIII
84	I
85	II

It is common, and it will be useful to you, to report the preceding data as 80–83, 82. This indicates the usual range and the best single figure to use. In fact, if you always report the usual minimum and maximum and the *average reading* you will never have to worry about getting it wrong.

Although at higher frequencies you may find the needle rock steady, resist the temptation to read to fractions of a decibel. That sort of precision just isn't there. If you report fractional decibels you mark yourself as a rookie. By the same token, ads with claims of "a 17.3 dB improvement" suggest that the people who measured and reported that amazing performance probably don't know much about acoustics or measurement.

INTERFERING NOISE

Sooner or later you will need to measure noise produced by a motor, let's say, while rocks are falling in a hopper nearby or a front end loader does its calisthenics. For this you set the meter to "fast" and read the lowest regular reading. This will only work for steady sources such as motors, fans, and gearboxes. Obviously the lowest reading you can find among the intermittent noise must belong to the steady noise.

Another technique is more obvious. This is to "bore in" on the source by getting the microphone within inches of it. It has two drawbacks. The first is that you will have the mike in that part of the near field where various parts of the source do not produce the same sound pressure level. Therefore, make such measurements at several locations. The second drawback is that if you bring the mike too close, the presence of the mike and meter will change the level somewhat. It is wise to limit "boring in" to about a 3 in. distance between the mike and the source.

WALK AWAY

For sources that dominate their area, watch the meter as you walk away from them. If you can find where the 6 dB with double distance starts and ends, you have some useful information. However, often you cannot find it. Instead you find that before a regular pattern of decreasing level with distance can be established, you are walking *into* the direct field of some other noise source.

Occasionally you find standing waves. The level goes up and down in a regular pattern as you walk away. This can only happen when pure tones are an important part of the noise. We've already seen that standing waves are caused by reflections so we are right on the track of curing the problem (by preventing reflection).

PEAKS ARE INTERESTING

Plot out your octave band data or at least visualize in your head what such a plot would show. Take an interest in any peaks in the curve. The sharper the peak, the more it should interest you. Quite sharp ones usually indicate things like scroll-type fans, gears, and similar equipment typified by a constantly recurring sequence of events.

Usually you will be able to hum or whistle a pure tone you can hear in that vicinity. You can even use a recorder (the medieval kind of flute, that is) or pitch pipe to pin down what musical note was nearest to it. This, if you don't mind looking a little odd on the job, gives you an excellent fix on frequency because you can look up the frequency of that note in a table.

PROBLEMS

1 Worked example of energy averaging. The assistant production manager makes a 10-min tour of his department several times a day. This is a significant part of his noise exposure. He spends very little time in fixed locations. In order to assign an average level to his tour, you walk with him. You note the noise level every 15–20 sec (use the 1-2-3 method). The tally shown below results. What level will you assign his tour?

Data and Initial Calculations

Level Found (dBA)	Number of Times It Was Found	Power	Product
82	1	1.6×10^8	1.6×10^8
83	0		0
84	3	2.5×10^8	7.5×10^8
85	1	3.2×10^8	3.2×10^8
86	2	4×10^8	8×10^8
87	0		0
88	3	6.3×10^8	18.9×10^8
89	6	8×10^8	48×10^8
90	4	10×10^8	40×10^8
91	1	12.5×10^8	12.5×10^8
92	2	16×10^8	32×10^8
93	4	20×10^8	80×10^8
94	8	25×10^8	200×10^8
95	2	32×10^8	64×10^8
96	0		0
97	1	50×10^8	50×10^8
98	1	63×10^8	63×10^8
Total times measured	39		
Total power			628.7×10^8

Method

(a) The data, as originally taken, are grouped by level.

(b) Multiply the equivalent power by the number of times it was found.

(c) Divide the sum of products by the number of samples to find the average power for each location.

(d) Convert this back to a level.

Calculations

(This has been partly done already.)

(c) $\dfrac{628.7 \times 10^8}{39} = 16.1 \times 10^8$

(**d**)   By calculator or "thumbnail log table," this is 92 dBA.

2   Your task is to assign a level to the job of adjusting a running envelope machine. The adjuster spends about 3 or 4 min pulling samples, inspecting them, and making adjustments. By holding the microphone in a position corresponding to where his ear was and being sure to spend about as much time as he does in each position, you come up with the following tally. What is the equivalent level he has experienced?

Data

Level (dBA)	Number of Times Found
92	5
93	2
94	2
95	3
96	6
97	2
98	3
99	1
100	0
101	2

Answer

97 dBA

6

ABSORPTION—THE GENERAL IDEA

〰〰〰〰〰〰〰〰

This is the quick universal symbol.

Of all the noise control techniques that exist, absorption must be the best known. In fact some people think it is the only noise control method. The first solution that occurs to the novice is to install an "acoustical ceiling." That often solves the problem, but sometimes it doesn't. Sometimes it solves the problem at a very high cost when another method of noise control might have been quite inexpensive. Occasionally an acoustical ceiling is just what you don't want! In a good sized meeting room where amplifiers are not used to aid the speaker, a reflective ceiling is a help—though absorption on the rear wall will probably be needed.

For locations near the source of noise, *absorption never works*. At least, it rarely does. Aside from that, it is more universally applicable to industrial noise control than any other method. It is no accident that we study it first. You will be able to solve a great many real problems with what you learn about it here.

It is not an inexpensive method but sometimes it is the cheapest way to solve the problem because other methods (enclosures, for example) carry other penalties such as restricted access and difficult maintenance.

As you work with absorption, always keep in mind that it is not an effective control method for locations near the source of noise. In that vicinity, the *direct* field of the source controls the sound pressure level. Absorption only works in the reverberant field.

WHAT IS ABSORPTION?

When a wavefront of sound strikes a large surface (large means that the dimensions of the surface are at least comparable to the wavelength of the sound), we expect it to be reflected. To the extent that it is not reflected it is absorbed. This definition of absorption is purposely broad because many mechanisms are capable of substantially reducing the energy reflected from a surface. In practice about three of them rate serious consideration. These are:

Porous Materials

Porous materials such as glass fiber, open cell foam, mineral wool, sintered metals, some unusually porous ceramics, and some sprayed coatings work because of two mechanisms that in an analogy to electrical theory are like resistance and reactance.

Reactive loss occurs by the delay in the transmission of pressure differences in the air—because of differences in path length and for more complicated reasons—so that the wavefront gets "confused" in traveling through these media and pressure peaks wander into rarefaction zones willy-nilly. Result? Much of the energy contained by the wave is dissipated.

The resistive mechanism works because the very fine glass fibers, for example, are made to move over each other by pressure difference. To have caused this the air must have had to give up the work represented by the force required to move them and the distance they moved. Since they dragged over each other, heat was generated by friction. That heat came from the energy of the sound wave. Hence less sound energy remained in the wave.

Panel Absorbers

Panel absorbers are resonant structures. They tend to be selective about the wavelengths (frequencies) they want to take out of circulation. It is possible to make them more broadminded by intelligent design.

They are the favorite of the acoustician who must balance a recording studio or performance hall. You can design them to eat up low frequency sound. You can do this with porous materials too—if you have a big enough budget!

Helmholtz Resonators

Helmholtz resonators also are selectively tuned to wavelength (frequency) and they are not generally useful as a technique of industrial noise control. There are two exceptions to that statement. A type of Helmholtz resonator in the form of a slotted hollow cement block is extremely useful in new construction. Outside the field of "absorption," Helmholtz resonators are useful as mufflers. The muffler on your car is a modification of the Helmholtz idea.

POROUS MATERIALS

Porous means porous. In fact in the good old days of acoustics the old China hands had a habit of testing any absorber that was new to them by holding it to their lips and blowing into it. In general terms the amount of pressure they had to bring to bear and the amount of air they expelled told them something about whether it was a good absorber or not. It was rather like horse trading or buying a used car—you kicked the tires a lot.

In this light what do you make of film covered foams, for example? Try the old horse trader's test. Is it an absorber? As a matter of fact it is. There is a porous foam under the thin film covering and wavefronts use up part of their energy in traveling through it. (Now this all assumes that it *is* a porous foam. Closed cell foams are usually poor absorbers with or without a film covering.) That the wavefront can get through the film covering will be left to the chapter on transmission loss ("Sufficient unto the day . . ."").

Old horse traders weren't as dumb as they might seem. A quantity known as *resistivity* (or specific resistivity) has been given official recognition as a useful quantity in judging absorbers by no less august body than ASTM. In fact, if you pursue the design of some mufflers and silencers (also a later chapter), you'll find you need to know how hard the horse trader blew and how much air he spent—all nicely wrapped up in "resistivity."

Porous absorbers are the first and mainstay sort of absorbers. Judged purely by the sort of absorption used—in square feet—they must easily account for over 95% of all absorbers that somebody paid for. The major exception is the slotted cinderblock Helmholtz resonator.

ALPHA

When sound waves strike a hard surface that is large enough they are reflected without loss. A "hard" surface, for example, Terrazo floor, vitrified tile walls, smooth, painted concrete or plaster, thick glass, is perfectly smooth and flat and cannot be moved.

If you represent the power of sound by W, then

$$W_r = W_i$$

where W_r = reflected power (any units)
W_i = incident power (same units)

Now if the surface is covered by one of the preferred porous materials some of the power is lost in the porous material and the amount of power reflected is less than the incident power. By how much? By the amount that was absorbed! Call the ratio of the reflected power to the amount of the power that was incident α. Then the earlier relationship changes to

$$W_r = W_i (1 - \alpha)$$

where W_r and W_i = reflected and incident power as before   (any units)
α = the absorption coefficient alpha   (no units)

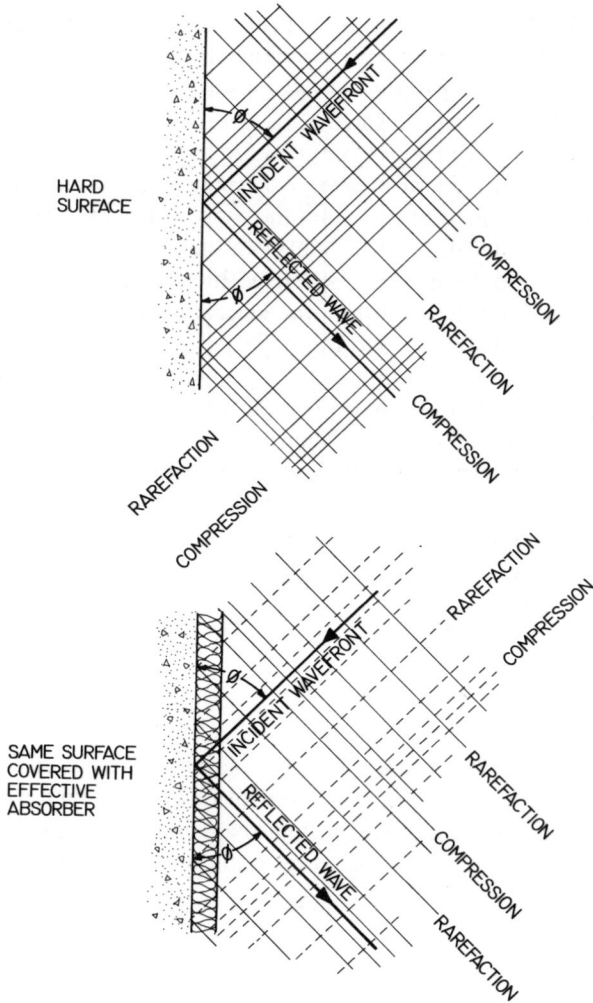

Figure 23 At a smooth, reflective surface the sound wave rebounds at the same angle as it hit. No energy is lost in the reflection process. If the surface is covered with an effective porous absorber (lower sketch), the same geometry holds but energy is lost in the process. Power in the reflected wave is less.

Be clear on this definition: Alpha is the *fraction* of sound power absorbed when the sound wave impinged once on the surface in question. Since it is a ratio, it needs no units and it will always have a value between 0 and 1.

Alpha is a simple fraction without units, but there is a unit of absorption. It is the *sabin*. A sabin is a square foot of perfect absorption ($\alpha = 1$). An open doorway measuring 3×7 ft has an absorption of 21 sabins. Ten square feet of material with $\alpha = 0.1$ has an absorption of 1 sabin.

You are not likely to run into metric sabins in the literature published in the United States—yet. They are used in Europe and are probably coming. A metric sabin, as you've already guessed, is a square meter of perfect absorption.

The usual abbreviation for absorption in sabins is A. (Be careful, though, A can also mean attenuation. In acoustics S is used as the symbol for area.)

In effect, then, alpha does have units. At least it is sometimes convenient to assign it the units sabins per square foot.

IMPOSSIBLE BUT TRUE

Alpha really should have a value between 0 and 1. It will seem painful at first to learn that α has been measured as 1.38 in a well-designed lab by a perfectly honest person, expert in making these measurements. An absorber that soaks up 138% of the sound power impinging on it? How can that be true?

Remember those expanding bubbles of sound waves (really air molecules jostling each other along in *all* directions)? Suppose you were one of those molecules. You are headed toward the edge of a very good absorber. It is on your right. On your left is a hard reflective surface.

You would notice as you traveled along at 1130 fps that there weren't as many molecules bouncing back toward you on the right as there were on the left, because there would be no reflected sound wave on your right. From the bustling scene on your left you could expect to be hit by a molecule sooner or later. That would nudge you toward your right *and toward the absorber.*

As an outside observer rather than a molecule, you could say that the sound wave has been bent into the absorber. The effect of this is a mixed bag. For you, the industrial noise control engineer, it is a blessing, because you will probably do a little better than theory and alpha say you should when you use absorption.

For the poor guy who ran the lab, alphas above 1 are a pain in the foot. Not only does he have to explain to an innocent public how the material absorbed more sound power than was there, he has to design his lab so that it is not fouled up by the effect. But that is another story.

Most of the absorption data you see follows a convention that when alpha is measured as more than 1 it is reported as 0.99. Some manufacturers do report the value actually measured, and the trend is to do this. Don't jump to the conclusion that there is any skullduggery connected with these practices. In both cases the lab or manufacturer is trying to tell you what to expect when you use the product they tested.

The mystery clears up when you consider two typical applications of absorption. The first is the use of a hung acoustical ceiling. If the ceiling area is 4000 ft^2, the usual hung ceiling will be about 4000 ft^2 also. It may be a bit less, because of flush lighting panels or ventilation openings, but generally the whole ceiling is absorptive. Contrast this with 1000 ft^2 of separated absorption panel units distributed over the same area.

The alpha for the acoustical ceiling will not exceed 1. The alpha for well-separated absorptive elements may well exceed 1. In general, the test reports have taken the probable use (continuous versus separated placement) into account. But there is a message in this practice, too. If you pack in absorber units with an alpha of 1.38 cheek by jowl, do not expect to realize better than 1.

When the lab test is made, 72 ft^2 of absorber are laid out in a compact area on the lab floor (typically). The reverberation room in which the test is made has a volume of 10,000 ft^3 or more. If the absorber has not been placed in a compact area, the lab report will point this out to you (though sometimes the catalog data may not). To be meaningful the test is set up so that the amount of sound power impinging on the absorber is about the same for all angles between the absorber and the path of the sound waves. This is called a diffuse field.

The bad lab is perhaps only 1000 ft^3 in volume. This means that the sound is mostly traveling *into* the absorbing surface—not at all diffuse. At worst the data may come from a "tube test" where all the energy is impinging at 90° to the surface.

Tube tests have their uses too—but they represent much less useful data to you as an industrial noise control engineer than good reverberation room data, (another story in acoustics that we must skip here).

Stay with this field and come to do some clever designing. It's fun. It beats bridge and crossword puzzles. Then you will wish that somebody would publish reliable data on alpha at various discrete angles of incidence. Such work has been possible since 1976—when the instrumentation was developed.

ENOUGH ROMANCE ABOUT ABSORPTION—NOW MEAT AND POTATOES

Here is a typical absorber. It is a glass fiber board 1 in. thick and weighing 2.25lb/ft^3. Porous absorbers are usually known by their density. The pounds per cubic foot are abbreviated pcf so this board is 1 in., 2.25 pcf.

First, notice that in Figure 24 the absorption coefficient alpha varies with frequency (wavelength). Considering how and why porous absorbers work this should be no surprise. In a meat-and-potatoes way it should tell you something different: when high frequency sound prevails (hissing, squealing, ringing) this absorber will be more useful to you than when the dominant sounds are at lower frequencies (booming, thudding, rumbling).

As Figure 25 shows, the difference in absorption with frequency is less pronounced as the thickness of the absorber is increased. Also note that thicker absorbers work better at low frequencies than thin ones. Another factor is the density of the absorber. The dashed line in Figure 25 is a reproduction of the curve of Figure 24 for 1 in., 2.25 pcf board of similar fiber size and bonding. The solid lines show the alpha of 3.25 pcf board of the thickness indicated.

Subtler tricks can be played. If you use the same material supported away from the reflecting surface, the low frequency absorption is increased and the

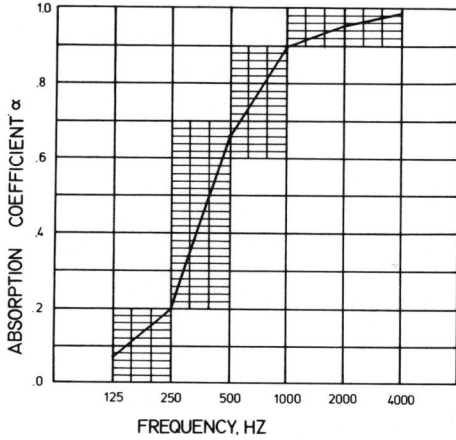

Figure 24 Absorption coefficients for a glass fiber board one inch thick and having a density of 2.25 pcf. The most notable feature is the great variation in alpha with frequency.

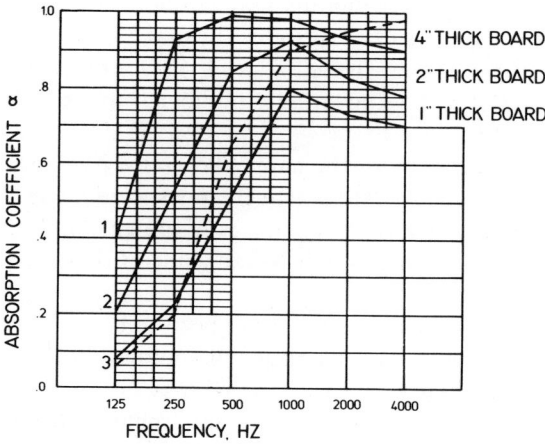

Figure 25 Solid curves are absorption coefficients for the three thicknesses shown of 3.25 pcf glass fiber board. The dashed line repeats the data of Figure 24. The absorption coefficient obviously varies with the thickness of the absorber. Considering the two curves for 1 in. thick material it also varies with the material—density being only one of the factors.

high frequency end of the curve is about the same. This is a clear gain acoustically, at the expense of space.

This trick of placing the absorber away from the reflecting surface has produced a standard test method known as a number 7 mount. In this case the absorber (typically a glass fiber board used for hung ceilings) is mounted in a track by conventional methods 16 in. from the reflecting surface. A good lab or catalog will always tell you when a number 7 mount has been used.

When no mounting is specified with the data you are using, assume that it was a number 4 mount. The specimen is laid on the concrete laboratory floor.

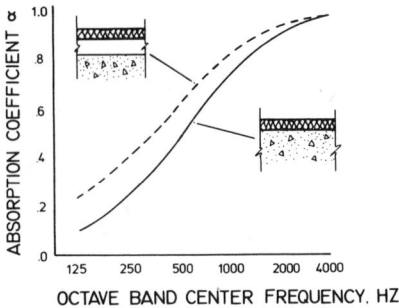

Figure 26 Spacing the absorber away from the reflecting surface it covers improves its performance at low frequencies.

The other possibilities are shown in Figure 27. Many of these are becoming uncommon and data for some of these mounts can only be found in older tests. As this book is being written, there is some ferment in the area of testing absorbers and a new set of standard test mounts may be in use soon.

If you plan to use an absorber in some unusual mounting, or if you want to see what would happen if you covered a lot of surface on ventilating ducts with absorber in what used to be known as a number 6 mount, you could ask the salesman. Good luck!

Instead, ask the salesman for the name and phone number of the director of research or the director of the lab where the tests were made. You will be surprised how humble and helpful that man is if you can find him. Many of the companies who support their own labs—and some of these are excellent—have an abundance of unpublished data.

If you wonder why these tests have never been publicized, you may be led to consider a marketing problem absorption manufacturers have. Think of the biggest room in your plant where you might want to use absorptive surfacing. You will probably have a list of special requirements for the material you employ. For example, you may need a grade of material that withstands oil mist or caustic soda fumes to cover two walls and the ceiling, at most.

Now think of the architectural market. An architect specified hung ceilings for the whole World Trade Center. Now honestly if you were a manufacturer of absorptive surfacing which of these two jobs would you go after? Well, the marketing departments of those companies have good sense too. The published data on commercially available absorbers are almost entirely tailored to an architect's problems.

The man at the top in research or testing will have a great deal of sympathy with you and an interest in what you are trying to do.

FILM COVERING

A second trick is shown in Figure 28. If the absorber is covered with a thin film covering, the alpha at low frequencies can be enhanced. This time there is a

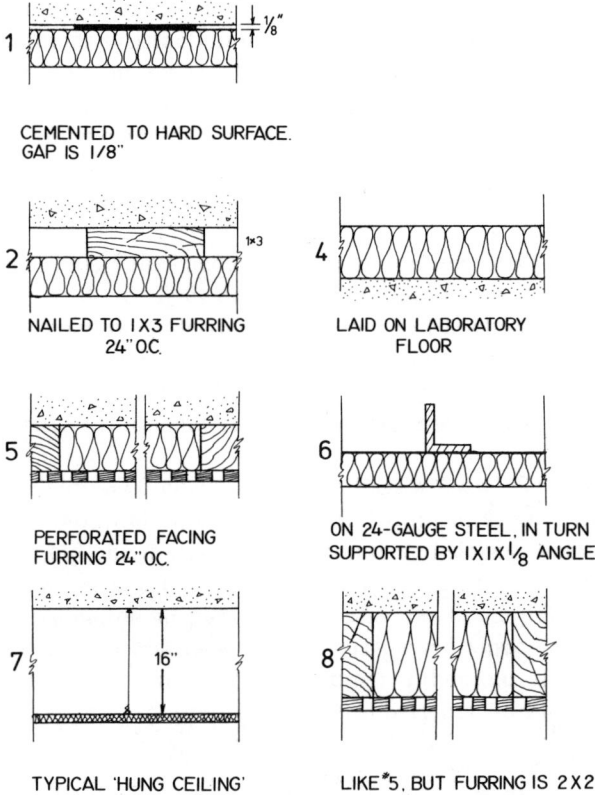

1

CEMENTED TO HARD SURFACE.
GAP IS 1/8"

2   NAILED TO 1X3 FURRING
           24" O.C.

4   LAID ON LABORATORY
              FLOOR

5   PERFORATED FACING
     FURRING 24" O.C.

6   ON 24-GAUGE STEEL, IN TURN
     SUPPORTED BY 1 X 1 X $\frac{1}{8}$ ANGLE

7   16"
     TYPICAL 'HUNG CEILING'

8   LIKE #5, BUT FURRING IS 2 X 2

STANDARD MOUNTING METHODS FOR TESTING
ABSORBERS

Figure 27 The six standard mounting methods used in testing absorbers. (Mounting number 3 is discontinued.) The number 4 and 7 mounts are the most commonly employed test methods.

price to pay. The performance gained at low frequency has been moved out of the high frequency performance. In some cases—furnaces, for example—this is still painless. Furnaces and boilers concentrate their noise energy in the low end and most any porous absorber, with or without a film facing, has performance to spare at the higher frequencies.

When you are using film coverings, think of 1 mil as the right place to set the film thickness. If you must still have high alphas at high frequencies, a 0.5 mil Mylar film is often a good choice. At 1 mil Mylar, polyvinyl chloride (available with Factory Mutual fire ratings) or polyethylene (lower cost and best weather resistance) are good choices. For film covering, about 1.5 mil is a benchmark. Films heavier than this begin to degrade performance of the absorber to the

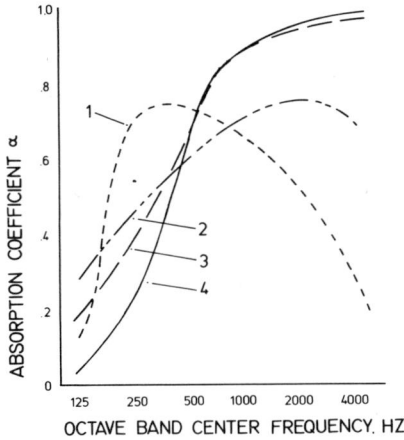

1 FACED WITH ADHERED 7 MIL FILM

2 BAGGED IN 7 MIL FILM

3 BAGGED IN LIGHT FILM (I MIL)

4 BARE ABSORBER

Figure 28 Film coverings can make useful improvements of low frequency absorption. As they become heavier, however, the high frequency performance suffers.

point where there is usually a less expensive way of achieving the result you want.

What result do you want? This discussion of modifying porous absorbers with a covering has so far imagined that the only question was good low frequency performance. [This is always available, by the way, if you simply use thicker layers of porous material. Compare the alpha at 125 Hz for 1, 2, and 4 in. material from the data (Table 6) as a project. Could you handle a problem of 32 Hz noise with a simple porous absorber? Of course you could—and the salesman who arranged it for you would probably give you a Shetland pony, or something!]

There are other reasons you may want to—or have to—cover a porous absorber. Some common ones are:

1 The operation throws an oil mist and has to be cleaned up or steamed down.
2 The operation involves a food product and any absorber must be non toxic, nonshedding, easily cleanable, and present no openings that harbor fungus or insects.
3 The operation may throw smoking hot, oil-covered metal chips.
4 Every once in a while, a log goes off the conveyor and crashes into the wall.

Surely you can imagine film coverings suitable for the first two situations. If you know about some for the latter two, you have some information the acoustical world has been waiting for!

Moral Sometimes you wish you didn't have to cover a porous absorber at all because you have a first interest in acoustic performance. However, there are other considerations like hot, abrasive chips or heavy objects moving fast and you must cover the absorber with something that will protect it.

FILM COVERINGS?

Try the hot abrasive chips first: Here a perforated metal covering with openings small enough to bar most of the chips may do nicely. As a project, find such a covering in Table 6 at the end of this chapter for chips with a minimum dimension of 3/16 in.

However, the original description of the problem said they were smoking hot, oil-covered chips. If they were thrown from a high speed machining operation oil mist or condensable oil vapor must be present, too. Is that constant loading of a *porous* absorber with oil going to change its alpha? To stop this from happening, let's put in a barrier to keep the oil that got through the holes out of the glass fiber board, and let's arrange to clean the oil from it.

Steaming down may do the trick. Now Mylar film between the perforated metal and the fiber board would solve the problem. Wasn't that easy? *No*—it has just fouled up the acoustics again!

It solves the oil mist problem and ruins the absorption of the glass fiber board. The general answer is to insert a "chair" that permits the sound waves to enter the openings in the perforated metal covering (and also do some tricks with the *unperforated* area of the metal sheet, as we shall see) and yet does not stop the film covering of the absorber from moving fairly freely. A typical "chair" in a vertical wall is a cheap wire fence material of number 10 wire in 4 in. squares.

For the other case (fast moving logs or lumber that might derail at 20 mph), you might unabashedly recommend a protective covering of 3 in. structural channels. Would you feel safe behind that barrier?

Where you have to resort to such heroic measures as 3 in. structural channel to keep material objects (lift trucks and leaning shoulders included) out of your absorber, you can read the number 4 mount data (or other mounts, if they apply) *without qualm as long as your protective barrier is at least 35% open.*

THE SIMPLE TRUTH

Sometimes, as you have just seen, that's how bad it can get using absorption. To be honest, however, these are, if not exceptional, at least "hard" cases. An easy case is a nice, dry room with no flying objects, oil mist, or anything that prevents you from using the absorbers produced for architects.

What can absorption do for you? Go back to fundamentals. It can absorb the sound energy that hits it instead of reflecting the sound. In terms of lowering noise level, absorption can do this:

$$NR = 10 \log\left(\frac{A_f}{A_s}\right)$$

where A_f = sabins of absorption after treatment
        A_s = sabins of absorption to start
     NR = noise reduction in decibels

Absorption works only in the reverberant field. When the noise source is distant and, for the most part, has to hit several reflecting surfaces before it makes its waves felt where you are, treating those surfaces will produce the noise reduction predicted by the equation.

SIMPLE EXAMPLE   A room (16 ft × 20 ft × 8 ft high) with about the same absorption at all surfaces (walls, ceiling, floor) has a sound pressure level of 93 dB. The average absorption coefficient ($\bar{\alpha}$) is 0.05 and the total surface area of the room is about 1200 ft². How much absorptive surfacing must be installed in this room to lower the L_p to 90 dB?

We know that the average alpha ($\bar{\alpha}$) is 0.05 and there are about 1200 ft² of room surface. We multiply 1200 ft² × 0.05 sabins/ft² to find the starting absorption in the room is 60 sabins.

Suppose we were to cover the 16 × 20 ft ceiling with an absorber whose alpha is 0.75. This would represent

$$16 \text{ ft} \times 20 \text{ ft} \times 0.75 \text{ sabins/ft}^2 = 240 \text{ sabins}$$

Thus the final absorption in the room would be 240 plus 60 sabins. According to the formula

$$NR \text{ (in dB)} = 10 \log\left(\frac{A_{final}}{A_{start}}\right)$$

we have

$$NR = 10 \log (300/60) = 10 \log 5 = 7 \text{ dB}$$

This, of course, is much more than is needed, so we could try covering part of the ceiling, or a wall, or any surface, and finally, by trial and error, arrive at the required final absorption of 120 sabins.

There are some errors in this simple approach and there are some shortcuts, elegant and otherwise. Consider the following.

Error   When you cover a surface with an absorber, you should deduct the alpha of the original surface from that of the absorber you are using. You are going to realize the *net* increase only. In this case, since the ceiling had an alpha of 0.05 and the absorber 0.75, you could estimate the absorption added as (16 × 20) (0.75 − 0.05) sabins.

Error   You may, indeed, find an average alpha for any frequency of interest by simply working out the total sabins in the room and dividing them by the total area. You will be wise not to try averaging alphas for a single material at several frequencies. Such a number does exist. It is not called $\bar{\alpha}$ but NRC (for noise reduction coefficient). Avoid NRC. Ignore it. It will get you in trouble unless you do. Work with alphas at known frequencies.

Here is the problem: Two absorbers may have the same NRC but one is good at the low frequency end and the other is good at the high frequency end. It now makes a great difference whether the noise source is a small furnace (nearly all low frequency) or, for example, a screaming circular saw. Surprisingly, if the source of noise has a spectrum that has about equal energy in every octave band, neither absorber will begin to live up to the promise made to you by its NRC.

Shortcut   If you are still handy with math, you will be able to juggle the NR $= 10 \log (A_f/A_s)$ equation so that you plug in the two knowns (NR and A_s) and solve directly for the required final absorption. At least it will give an idea of the quality and quantity of absorber to try.

$$A_f = A_s \times 10^{(NR/10)}$$

Shortcut   If you want to achieve the same result without wracking your memory about logs and exponential equations, you can use the nomograph of Figure 29.

Practical Tip   When it doesn't matter which surface you cover with absorption, put it in corners of the room. Absorption is most effective there because, to put it simply, sound energy tends to concentrate in corners.

Even if you can't use the surfaces in the corners, it is better practice to use the required amount of absorptive surfacing distributed in the three planes of the room surfaces rather than to lump it all in one. A common misapplication of absorptive surfacing in architectural acoustics, for example, is to put in a hung ceiling in a deeply carpeted room.

Sound waves bouncing vertically in the room have already been absorbed in good degree by the carpet and a smaller amount of absorption on the walls would accomplish much more than absorption on the ceiling could.

Porous absorbers, as has been noted, are the usual choice for industrial noise control. Here we've spent most of our time looking at them, particularly at glass fiber board. Everything discussed applies directly to glass fiber batts (limp, easily deformed panels) and blankets. The real difference between "batts" and "blankets" is size.

Batts are typically 2 $\times$ 4 ft by whatever thickness in inches you order. Blankets come in long rolls. For the small job you may want to visit the local lumber yard and pick up homebuilder's insulation—typically about 15 in. wide and furnished in rolls of several feet. No need to warn you that you don't want foil or paper-covered glass (those are *thick* films!) and that if you must use a

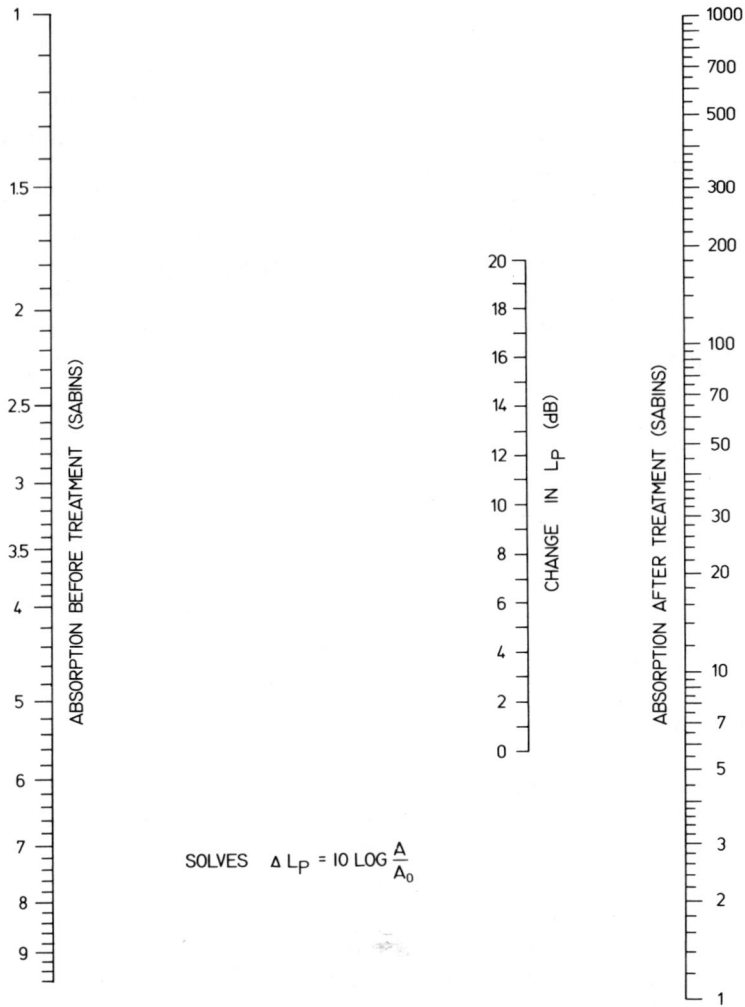

Figure 29 This nomograph performs the noise reduction calculation for absorption directly. To use it, estimate the absorption originally present in the room and locate it on the left-hand scale. For example, if the starting absorption for some frequency is 350 sabins, locate 3.5 on the left scale. Both the left-hand and right-hand scales should now be read in *hundreds*. Thus if the final absorption is to be 500 sabins align your straightedge on 3.5 at left and 5 at right and read a change in L_p (noise reduction) of about 1.5 dB. On the other hand, if the final absorption is to be 5000, align 3.5 and 50 to read a reduction of 11 dB.

paper covered material the paper should go against the surface you're covering so that the glass is exposed to the sound.

Mineral wool, open-cell foam (typically 2 pcf, open-celled, soft urethane), film covered foam, some sprayed coatings, and various other specialties are also porous absorbers and behave like the glass fiber boards described in this chapter. They deserve more attention than we've given them. They may offer you lower cost, better durability, or greater ease of installation than glass fiber board. Look into them by all means. They have been given little attention here simply because they behave in the ways described.

PROBLEMS

1 Worked problem. Off the large main room of a manufacturing plant there is a 10 × 10 × 8 ft high alcove. It houses a hydraulic pump. The evident source of noise in the room is the pump.

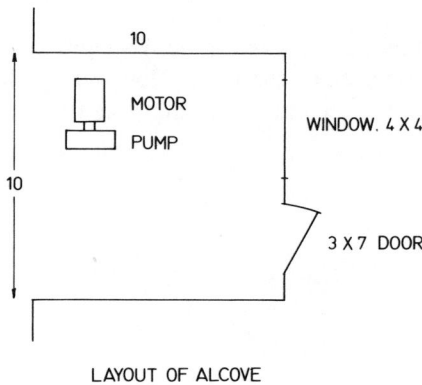

LAYOUT OF ALCOVE

Data

Sound pressure levels are taken at several locations in the room and the following averaged data are obtained:

Frequency (octave band center)	Sound Pressure Level (dB)
125	97
250	101
500	108
1000	115
2000	112
4000	106

The finish and normal contents of the room are:

Floors and ceilings, poured concrete.
Three walls, painted brick.
Fourth wall, open to plant.
Miscellaneous, pump, motor, piping and wiring in conduit, a 4 × 4 ft window of 1/8 in. glass, a 3 × 7 ft metal exterior door (both in the wall opposite the opening). The room normally contains two 55-gal drums partly filled with trash and one or two light crates or corrugated cartons.

Part 1 Do you have a problem in this space if any employee spends any time in it? What is the A-weight level?

Method

The best idea is to go there and measure the A-weighted level, because you will probably learn much more than the level. However, if the plant is 500 miles away or your information is second hand, compute the level in the normal way.

Calculation

Two methods, difference table and short log table, are shown.

Frequency (Hz)	Octave Band L_p (dB)	A-Weight Correction (dB)	Effective (Corrected) Level (dB)	Short Log Table (power referenced to 100 dB) (equivalent to 1 unit)
125	97	−16	81	0
250	101	−9	92	0.16
500	108	−3	105	3.16
1000	115	0	115	31.6
2000	112	1	113	20
4000	106	1	107	6.3
Total power =				60

Total power is approximately equal to 60 units, and this is closest to 63 units in the table, equivalent to 118 dBA.

Calculation by Difference Table

Frequency	Effective (corrected) level
125	81
250	92
500	105
1000	115
2000	113
4000	107

105 ⟶ 109
115
113 ⟶ 114.5
⟶ 118 dBA

If you used a calculator, you found 117.78 dBA. This, of course, is also 118 dBA.

Answer

Indeed you are in trouble if people spend any time in this room. The highest level permitted for any exposure is 115 dBA. The room is at 118 dBA.

Part 2 What is the *first* thing you should do about correcting this situation?

(a) Build a wall to close off the alcove from the plant?

(b) Find out whether the pump is in good repair?

(c) Use a stethoscope to find out where the noise originates?

(d) Ask the foreman, a veteran of 14 years in the plant, whether it has always been this noisy

(e) Add absorption to the alcove surfaces?

Answers

Answers (b), (c), and (d) are all reasonable *first* choices. If you chose (a) you should really stop and cogitate for a moment. The problem is noise in the alcove and closing in the open wall will raise the noise level there. (The open "wall" is perfect absorption and closing it with *anything* will raise the level because the wall will be less than perfect absorption.) If you chose (e) you can hardly be faulted because this is a book about noise control and a chapter about absorption. If all this had happened to you outside a book, wouldn't you have tried (b), (c), and (d) first?

Part 3 What noise reduction can you achieve in this alcove if you line the two featureless walls and the ceiling with 1 in. 3 pcf glass fiber board in a number 4 mount?

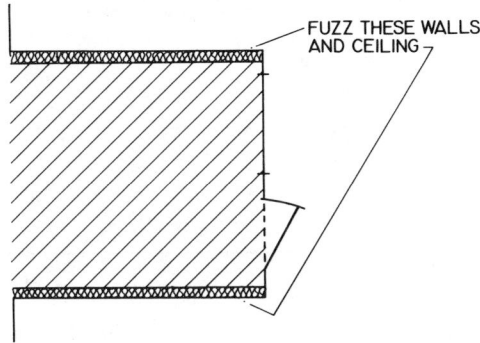

Method

(a) Make a table in which you list all the items that offer absorption in the room *as is* and set up a column in your table for each octave band. It will be convenient to tabulate both the alpha and sabins in each pigeon hole of your table (see table below if this is not clear).

(b) Sum the sabins of absorption for each frequency for the room *as is*.

(c) Repeat (a) and (b) for the room after you have added absorption.

(d) Find the noise reduction for each frequency by NR $= 10 \log(A_f/A_s)$.

(e) Apply the noise reduction to the original data at each octave.

(f) Find the new A-weight.

Calculation

(a, b) Find total absorption for the *as is* (A_s) alcove. *Note* the absorption coefficients for some common building materials are given in Table 5 at the end of this chapter.

Item (of absorption)	Area (ft^2)	**Absorption Coefficient, α, at Frequency (Hz)** Absorption, sabins, at Frequency (Hz)					
		125	200	500	1000	2000	4000
Floor and ceiling	200	**0.01**	**0.01**	**0.01**	**0.02**	**0.02**	**0.02**
(poured concrete)		2	2	2	4	4	4
Three walls less door	203	**0.01**	**0.01**	**0.02**	**0.02**	**0.02**	**0.02**
and window		2	2	4	4	4	4
(painted brick)							
Window (1/8 in.	16	**0.35**	**0.25**	**0.18**	**0.12**	**0.07**	**0.04**
glass)		6	4	3	2	1	1
Door (assume 18	21	**0.2**	**0.1**	**0.05**	**0.02**	**0.02**	**0.02**
gauge steel)		4	2	1	0	0	0
Open side of alcove	80	**1**	**1**	**1**	**1**	**1**	**1**
		80	80	80	80	80	80
Miscellaneous		10	10	10	10	10	10
(assume 10 sabins)							
Total absorption, in sabins, to start		104	100	100	100	99	99

(c) Find total absorption in the alcove after treatment (A_{final}).

| | | **Absorption Coefficient, α, at Frequency (Hz)** | | | | | |
| | | Absorption, sabins, at Frequency (Hz) | | | | | |
Item (of absorption)	Area (ft^2)	125	200	500	1000	2000	4000
Floor only*	100	1	1	1	2	2	2
Ceiling (1 in. 3 pcf)	100	0.03	0.22	0.69	0.91	0.96	0.99
		3	22	69	91	96	99
Less original ceiling*		−1	−1	−1	−2	−2	−2
Two end walls	160	0.03	0.22	0.69	0.91	0.96	0.99
		5	35	110	146	151	158
(Less original walls)*		−2	−2	−3	−3	−3	−3
Third wall less	43	0.01	0.01	0.02	0.02	0.02	0.02
window and door		0	0	1	1	1	1
Window (from above)		6	4	3	2	1	1
Door (from above)		4	2	1	0	0	0
Miscellaneous (from above)		10	10	10	10	10	10
Open side (from above)		80	80	80	80	80	80
Total, final (sabins)		106	151	271	327	336	346

*from table just above

(d) Find the noise reduction in each octave.

Octave	Ratio A_f/A_s	10 log Ratio $=$ NR (dB)
125	1.02	0
250	1.51	2
500	2.71	4
1000	3.27	5
2000	3.39	5
4000	3.49	5

(e) Apply this noise reduction in each octave.

Original level (corrected for A-weight, see Part 1)	81	92	105	115	113	107
Less noise reduction	0	2	4	5	5	5
New octave band level after adding absorption	81	90	101	110	108	102

(f) The new A-weighted level, by the familiar methods shown in part *a* of the method is 113 dBA. Thus, adding this absorption to the alcove appears to have achieved a 5 dBA noise reduction. Has it really? In all parts of the alcove? No, because most of the alcove is in the direct field of the source and absorption reduces the level in the reverberant field. The before and after sketches here are typical of what you might expect. Depending on the positions used to find an average level in the alcove that average may have dropped two or three dBA.

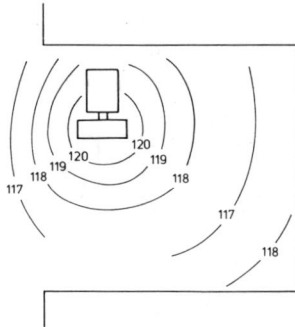

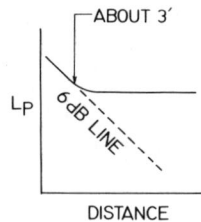

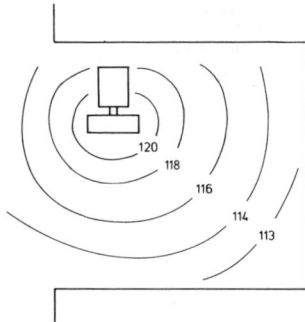

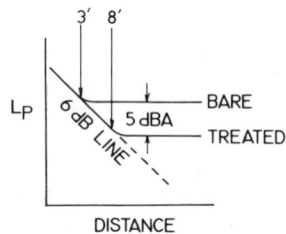

2   Worked problem. Suppose the alcove for the first problem had the same sound pressure levels in each octave but that the source of noise was far outside the alcove. This changes the situation in two important ways: First, since you are not in the direct field you can expect absorption to do a better job. Second, all the sound power is coming through the open end of the alcove so this is not a symmetrical space. Is there a way of predicting what will happen when absorption is added to a space like this?

Method

(a) In some situations you may want to split the space up into several sections depending on geometry or other factors. The ultimate attack on this sort of problem might even involve integrating along surfaces. Even if math comes painlessly to you do not expect good, useful results in the typical "dirty" industrial noise control problem from such elegant methods—save them for the nice "clean" case of estimating when to stop using an absorptive lining in a tunnel or circular duct.

In this problem it will suffice to split the room into the two obvious parts. Draw a line midway between the opening to the alcove and the reflective wall that faces that opening. Make up two new rooms by flipping the left and right halves of the existing room over to form their mirror images. The two new hypothetical rooms you have created will be bordered on all sides by walls ("best" room) and open on two sides ("worst" room).

(b) Exactly as before estimate the absorption in each of these rooms before and after treating them with absorption. The calculation shown here only summarizes the results (areas and alphas are omitted from the table). It is suggested that you make a complete table and work through the problem using these summary intermediate results to check your understanding of the method.

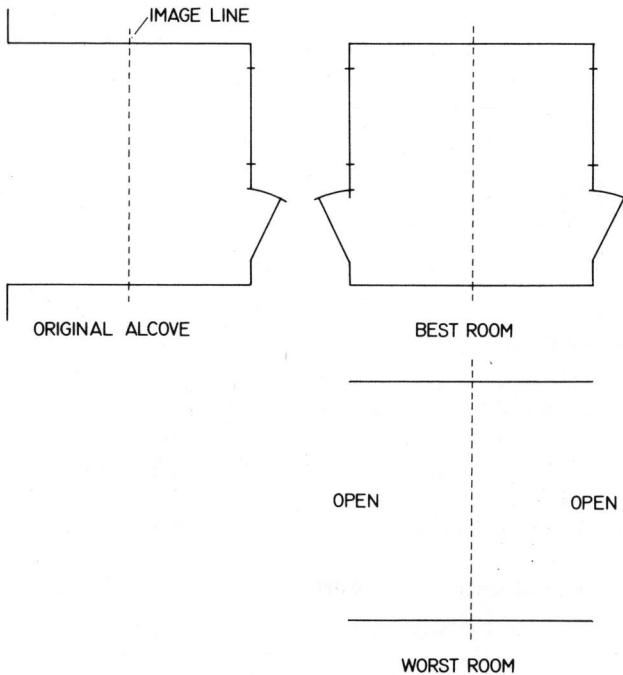

Also note that *all* the assumed miscellaneous absorption (10 sabins) has been assigned to the "best" room. It is much more critical in that space than in the "worst" room and it is conservative practice to assign it to the more reverberant space.

(c) Go back to the original plan of the alcove and lay in your results for each half in the center of that half. If necessary, sketch in contour lines of the expected A-weighted sound pressure level.

Calculation

(a) "Best" room absorption, to start (in sabins):

| | Frequency (Hz) | | | | | |
Item	125	250	500	1000	2000	4000
Floor and ceiling	2	2	2	4	4	4
Walls less window and door	2	2	5	5	5	5
Two windows	11	8	6	4	2	1
Two doors	8	4	2	1	1	1
Miscellaneous	10	10	10	10	10	10
Total, to start	33	26	25	24	22	21

(b) "Best" room absorption, final:

| | Frequency (Hz) | | | | | |
Item	125	250	500	1000	2000	4000
Floor	1	1	1	2	2	2
Ceiling, net	2	21	68	89	94	97
Two end walls, net	3	33	107	143	150	155
Two side walls	1	1	2	2	2	2
Two windows	11	8	6	4	2	1
Two doors	8	4	2	1	1	1
Miscellaneous	10	10	10	10	10	10
Total, final	36	78	196	251	261	268
NR $= 10 \log(A_f/A_s)$	0	5	9	10	11	11 dB
"Best" room original level, corrected to A-weight	81	92	105	115	113	107
Same, less NR	81	87	96	105	102	96

The final A-weight level for the "best" room is 107 dBA.

(c) "Worst" room absorption, to start:

| | Frequency (Hz) | | | | | |
Item	125	250	500	1000	2000	4000
Floor and ceiling	2	2	2	4	4	4
Two end walls	2	2	3	3	3	3
Two open "walls"	160	160	160	160	160	160
Total, to start	164	164	165	167	167	167

(d) "Worst" room absorption, final:

Item	Frequency (Hz)					
	125	250	500	1000	2000	4000
Floor	1	1	1	2	2	2
Ceiling, net	2	21	68	89	94	97
Two end walls, net	3	33	107	143	151	155
Two open "walls"	160	160	160	160	160	160
Total, final	166	215	336	394	407	414
NR = $10 \log(A_f/A_s)$ =	0	1	3	4	4	4
"Worst" room original level, corrected to A-weight	81	92	105	115	113	107
Same, less NR	81	91	102	111	109	103

The final A-weight level for the "worst" room is 114 dBA.

Answer

Now the results actually achieved will not only lie between these extremes but the two parts of the room will tend to have the values calculated. The contours you might expect are shown here.

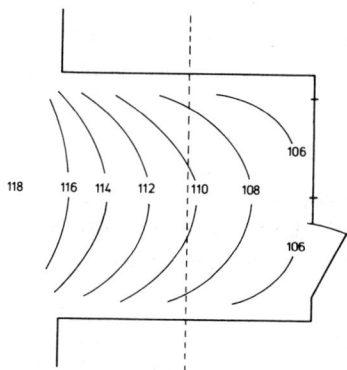

3 Partially worked problem. Off the main floor of a plant is a work room 20 × 10 and 8 ft high. It has a bandsaw whose scream is the only important noise source in the room.

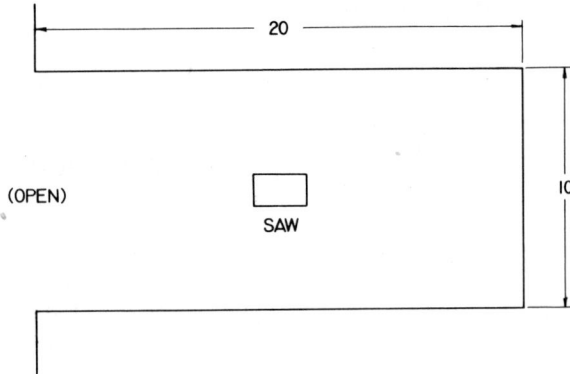

Data

The bandsaw has three teeth per inch and is running on 30 in. diameter wheels that turn at 200 rpm. It is cutting corrugated cardboard.

The finish of the room is:

Floor, 10 × 20 ft painted concrete.

Ceiling, 10 × 20 ft poured concrete, rough unpainted.

Two long walls, 8 × 20 ft painted brick.

Short wall, 8 × 10 ft painted plaster, smooth.

Miscellaneous, 200 ft^2 exposed stacks of corrugated cardboard (assume alpha at 1000 Hz is 0.15); 150 square feet of sheet metal ducting and dust collector (say alpha at 1000 Hz is 0.1).

Part 1 (Worked)   What frequency is the scream? If you had an octave or one-third octave band meter you would measure it, of course. Suppose you hadn't?

Method

The scream is produced by impact of the bandsaw blade teeth. The frequency at which they hit can be calculated. It will probably be the worst frequency although the harmonics may also be important.

$$\frac{3 \text{ teeth}}{\text{in.}} \times 30 \text{ in.} \times \pi \times \frac{200 \times \min}{\min \times 60 \text{ sec}} = 942 \text{ Hz}$$

Part 2   Can you help the saw operator by absorptive treatment of this room?

Part 3   The operator's helper is found to be exposed to 100 dBA around this operation even though he or she spends no time in the direct field of the saw.

Can you find a room treatment that will lower the exposure level to 97 dBA or less? Base your estimate on the pure tone only and assume that the saw is the only source. Use the simple calculation (ignore the asymmetry of the space). You may use any absorber and cover any surface except the floor.

Part 4 Because of the dust problem, management is considering closing off this alcove. Do a quick, simple estimate of the change in reverberant noise level if the opening of the bare alcove were walled off with $\frac{1}{4}$ in. fir plywood.

Answers

Part 2 You did say "No," didn't you? Direct field!

Part 3 The room had about 145 to 160 sabins to start (don't forget the two people and the open wall). Covering the ceiling or one long wall with 1 in. 3 pcf glass fiber board would achieve about 3 dBA noise reduction even if the harmonics are prominent in the spectrum because it has a good alpha at higher frequencies. (You can base your calculation on the single octave band centered on 1000 Hz.)

Part 4 The decrease in absorption will increase the reverberant noise about 3 dBA (based on the 1000 Hz octave). Use the familiar $NR = 10 \log(A_f/A_s)$ equation.

Table 5 *Absorption Coefficient of Common Materials*

Thickness (in.)	Material	Absorption Coefficient (octave band center frequency)					
		125	250	500	1000	2000	4000
Part 1	**Common Building Materials**						
18	Brick wall, bare	0.025	0.025	0.03	0.04	0.05	0.07
18	Brick wall, painted	0.01	0.01	0.02	0.02	0.02	0.02
18	Brick wall, whitewashed	0.02	0.02	0.02	0.03	0.03	—
6–12	Concrete block, coarse	0.36	0.44	0.31	0.29	0.39	0.25
6–12	Concrete block, painted	0.10	0.05	0.06	0.07	0.09	0.08
	Concrete, poured, bare	0.01	0.01	0.02	0.02	0.02	0.03
			to	to	to	to	to
			0.02	0.04	0.06	0.08	0.10
	Concrete, poured, oil paint	0.01	0.01	0.01	0.02	0.02	0.02
	Concrete, poured, painted and varnished	0.01	0.01	0.01	0.02	0.02	0.02
	Concrete, rendered	0.004	0.004	0.005	0.006	0.008	0.015
$3/4$	Cork slab, glued down	0.08	0.02	0.08	0.19	0.21	0.22

Table 5 *Absorption Coefficient of Common Materials*

Thickness (in.)	Material	Absorption Coefficient (octave band center frequency)					
		125	250	500	1000	2000	4000
3/4	Cork slab (as above) waxed and polished	0.04	0.03	0.05	0.11	0.07	0.02
—	Floor tile on concrete	0.02	0.03	0.03	0.03	0.03	0.02
—	Floor tile on subfloor	0.02	0.04	0.05	0.05	0.10	0.02
—	Floor tile, rubber, on concrete	0.019	0.033	0.04	0.036	0.018	0.02
3/16	Floor tile, linoleum, on concrete	0.04	0.03	0.04	0.04	0.03	0.02
3/16	Floor tile, vinyl, on concrete	0.04	0.03	0.04	0.04	0.03	0.02
1/4	Glass, small area	0.04	0.04	0.03	0.03	0.02	0.02
1/4	Glass, large area	0.18	0.06	0.04	0.03	0.02	0.018
1/8	Glass, single weight window	0.35	0.25	0.18	0.12	0.07	0.04
1/2	Gypsum board on 2 × 4 studs 16 in. o.c.	0.29	0.10	0.05	0.04	0.07	0.09
—	Linoleum on concrete	0.01	0.01	0.02	0.02	0.03	0.03
1/16	Molded FRP, glass fiber reinforced plastic, large area	0.48	0.25	0.11	0.04	0.04	0.04
1/16	Molded FRP (as above) with 1/32 in. undercoating	0.27	0.22	0.11	0.04	0.04	0.04
1/16	Molded FRP (as above) with 1/8 in. undercoating	0.14	0.17	0.13	0.04	0.04	0.04
2	Plaster on metal lath	0.08	0.06	0.05	0.04	0.04	0.04
2	Plaster, fibrous	0.35	0.30	0.20	0.55	0.10	0.04
—	Plaster, gypsum	0.01 to 0.04	0.01 to 0.04	0.02 to 0.04	0.03 to 0.06	0.04 to 0.06	0.03 to 0.06
3/4	Plaster, lime, sand finish on metal lath	0.04	0.05	0.06	0.08	0.04	0.06
3/8	Plywood over 1 in. air space	0.28	0.22	0.17	0.09	0.10	0.11
1/4	Plywood over 4 in. glass fiber	0.30	0.11	0.06	0.05	0.03	0.02
1/8	Plywood over 2 in. air space	0.11	0.21	0.10	0.05	0.03	0.02
—	Slate, solid backing	0.01	0.01	0.01	0.02	0.02	0.02
—	Terrazo	0.01	0.01	0.015	0.02	0.02	0.02
—	Wood, solid backing	0.04	0.04	0.03	0.03	0.03	0.02

Table 5 *Absorption Coefficient of Common Materials*

Thickness (in.)	Material	Absorption Coefficient (octave band center frequency)					
		125	250	500	1000	2000	4000
3/4	Wood, pine sheathing	0.10	0.11	0.10	0.08	0.08	0.11
—	Wood over deep air space (test made on wood platform)	0.40	0.30	0.20	0.17	0.15	0.10
3/8–1/2	Wood paneling over 2 to 4 in. air space	0.30	0.25	0.20	0.17	0.15	0.10
—	Wood, varnished floor on wood beams	0.15	0.11	0.10	0.07	0.06	0.07
—	Wood block floor	0.05	0.03	0.06	0.09	0.10	0.22

Part 2   Room Finish Materials
Carpeting

Thickness (in.)	Material	125	250	500	1000	2000	4000
—	Heavy, on concrete	0.02	0.06	0.14	0.37	0.60	0.66
	Heavy, on 40 oz pad	0.08	0.22	0.55	0.69	0.72	0.75
7/16	Amritza on concrete	0.09	0.06	0.24	0.24	0.24	0.11
	Axminster	0.11	0.14	0.20	0.33	0.52	0.82
7/16	Cardinal Batala on concrete	0.12	0.10	0.28	0.42	0.21	0.33
3/8	Pile, on concrete	0.09	0.08	0.21	0.26	0.27	0.37
5/16	Pile, on 1/8 in. felt	0.11	0.14	0.37	0.43	0.27	0.25
3/16	Rubber-backed, on concrete	0.04	0.04	0.08	0.12	0.03	0.10
—	Tufted nylon on 1/4 in. foam	0.06	0.14	0.34	0.36	0.31	0.37
3/8	Wool pile on concrete	0.09	0.08	0.21	0.26	0.27	0.37
5/8	Wool pile on pad	0.20	0.25	0.35	0.40	0.50	0.75

Drapery

	Material	125	250	500	1000	2000	4000
	Cotton, 14 oz/yd.^{2} flat at wall	0.04	(0.1)	0.13	(0.3)	0.32	(0.3)
	Cotton (same) draped to seven-eights of its area	0.03	0.12	0.15	0.27	0.37	0.42
	Cotton (same) draped to three-fourths of its area	0.04	0.23	0.40	0.57	0.53	0.40
	Cotton (same) draped to one-half of its area	0.07	0.31	0.49	0.81	0.66	0.54
	Velour, 10 oz/yd.^{2} flat at wall	0.03	0.04	0.11	0.17	0.24	0.35
	Velour, 14 oz/yd.^{2} flat at wall	0.05	0.07	0.13	0.22	0.32	0.35
	Velour (same) draped to one-half of its area	0.07	0.31	0.49	0.75	0.70	0.60

Table 5 *Absorption Coefficient of Common Materials*

Thickness (in.)	Material	Absorption Coefficient (octave band center frequency)					
		125	250	500	1000	2000	4000
	Velour, 18 oz/yd.^{2} flat at wall	0.05	0.12	0.35	0.45	0.38	0.36
	Velour, (same) draped to one-half of its area	0.14	0.35	0.55	0.72	0.70	0.65
Glass fiber board finishes (see Table 6)							
Mineral wool finishes (values are representative, performance varies)							
1	6 to 9 pcf, number 4 mount	0.06	0.19	0.39	0.54	0.59	0.75
4	(same)	0.42	0.66	0.73	0.74	0.76	0.79
Part 3	**Miscellaneous**						
3	Ashes, loose dumped (2.5 lb water/ft^3)	0.25	0.35	0.65	0.80	0.80	—
11	Ashes (same)	0.90	0.90	0.75	0.80	—	—
4	Gravel soil, loose and moist	0.25	0.60	0.65	0.70	0.75	0.80
12	Gravel soil (same)	0.50	0.65	0.65	0.80	0.80	0.75
—	People (values are sabins per warm body)	2.5	3.5	4.2	4.6	5.0	5.0
4	Sand, sharp and dry	0.15	0.35	0.40	0.50	0.55	0.80
12	Sand (same)	0.20	0.30	0.40	0.50	0.60	0.75
4	Sand (14 lb water/ft^3)	0.05	0.05	0.05	0.05	0.05	0.15
—	Water surface	(0)	0.01	0.01	0.01	0.01	0.02

Table 6 *Absorption Coefficients of Glass Fiber Board and Blanket Materials*[1]

Material[2]	Thickness (in.)	Mount or construction	Absorption Coefficient Frequency (Hz)					
			125	250	500	1000	2000	4000
1 in. Thick Treatments								
701	1	4	0.12	0.28	0.73	0.89	0.92	0.93
703	1	4	0.03	0.22	0.69	0.91	0.96	0.99
705	1	4	0.08	0.25	0.74	0.95	0.97	0.99
Linear	1	4	0.03	0.17	0.63	0.87	0.96	0.96
TIW	1	4	0.11	0.33	0.70	0.80	0.86	0.85
703	1 in., ⅛ in. pegboard cover	4	0.09	0.35	0.99	0.58	0.24	0.10
703 FRK	1	4	0.12	0.74	0.72	0.68	0.53	0.24
2 in. Thick Treatments								
701	2	4	0.24	0.77	0.99	0.99	0.99	0.99
703	2	4	0.22	0.82	0.99	0.99	0.99	0.99

Table 6 *Absorption Coefficients of Glass Fiber Board and Blanket Materials*[1]

Material[2]	Thickness (in.)	Mount or construction	Absorption Coefficient Frequency (Hz)					
			125	250	500	1000	2000	4000
705	2	4	0.19	0.74	0.99	0.99	0.99	0.99
703	2, perf. metal cover	4	0.18	0.73	0.99	0.99	0.97	0.93
TIW	2	4	0.25	0.75	0.99	0.99	0.99	0.99
Linear	1 in. over 1 in. air space		0.04	0.26	0.78	0.99	0.99	0.98
4 in. Thick Treatments								
701	4	4	0.73	0.99	0.99	0.99	0.99	0.97
703	4	4	0.84	0.99	0.99	0.99	0.99	0.97
TIW	4	4	0.57	0.99	0.99	0.99	0.99	0.99
Linear	1 in. over 3 in. air space		0.19	0.53	0.99	0.99	0.92	0.99
Number 7 mount								
701	1	7	0.56	0.85	0.70	0.89	0.93	0.99
703	1	7	0.65	0.94	0.76	0.98	0.99	0.99
703 FRK	1	7	0.48	0.60	0.80	0.82	0.52	0.35

[1]Reproduced by permission of Owens-Corning Fiberglas Corporation. Since their product code is shorter than a generic description of each sample, the table retains their product designation. The code is given just below. *Caution*—the performance of other products of the same generic description can vary widely from those given here! Fiber diameter, degree of bonding, and homogeneity of density, among other factors, are responsible for this.

Note that these data apply to large areas. Recent published data on these products show α greater than 1 and are applicable where some fraction of the surface is to be covered (see "impossible but true" section).

[2]*Code to Table*

Boards

701—bonded glass fiber, density 1.5 pcf

703—bonded glass fiber, density 3 pcf

705—bonded glass fiber, density 6 pcf

703 FRK—same as 703 but faced with foil/scrim/kraft paper laminate and tested with foil side toward the sound.

Linear—4.5 pcf board faced with bonded glass fiber cloth.

Blankets

TIW—thermal insulating wool, unbonded, 1 pcf density.

7

USING ABSORPTION AND SPECIAL ABSORBERS

FUNCTIONAL ABSORBERS

There is one slightly different porous absorber. It may be made of any one of several of the *materials* that fall in the porous absorber class. What sets it apart is the fact that it is not used to cover a surface but, instead, hangs free in the room or space that needs more absorption.

Its first attraction is that it is usually the least expensive way to add absorption to a room. The units (typically 2 × 4 ft. × 1 or 2 in. thick) are simply hung like wash in the waste space overhead. Since these absorbers don't cover any surface, they do not detract from the absorption already in the room. Functional absorbers are widely used.

They do have some drawbacks, however. In usual sizes, they are too small to have dimensions approximately the same as a wavelength for low frequency noise. Thus they do not offer much in cases where most of the energy is at low frequency. To make the best of a bad situation, they are usually film covered. A cylindrical type (typically 2 ft × 1 ft diameter) may even have built-in Helmholtz resonators to help with the low frequency.

They are also most effective when used, if not sparingly, at least not lavishly. Too many functional absorbers in the same space compete strongly with each other for the available sound energy and cannot live up to the expectations of a simple calculation of the noise reduction they should produce.

A good rule of thumb is that they should be separated by *at least* their smaller dimension to be reasonably effective.

PANEL ABSORBERS

To put things in perspective, suppose you were now faced with the problem of quieting a reverberant room housing a big reciprocating compressor (peak noise at 32 Hz). Most porous absorber data give 125 Hz as the lower frequency limit. Thirty-two hertz is two octaves below the last known datum. You could try sketching the curves and extrapolating alpha at this frequency or phone for help. In the end, porous absorbers would look unattractive.

After looking into what could be done by spacing the absorbers away from the reflecting surface and covering them with film, you would come to the conclusion that both the material cost and space required (a few feet) for the absorbing layer were unattractive.

This might be a good place to consider a panel absorber. These structures are panels thin enough to flex easily and large in extent that are framed along the edges so that the panel is some fixed distance from the reflecting surface.

The frequency that such a panel absorbs best is given by

$$f = 170/\sqrt{Wd}$$

where $f =$ the frequency (Hz)
   $W =$ the weight of the panel (psf)
   $d =$ the distance between panel and reflecting surface (in.)

One problem with a simple panel absorber is the sharpness with which it is tuned. Variations in a particular lot of plywood (if that were the panel material chosen) would be enough so that several "identical" panels would have measurably different peak absorption frequencies. Unless you know exactly what frequency you are trying to absorb and are willing to tune each panel to match it, this won't be an easy job.

This is one reason that panel absorbers are not often used. The main reason, of course, is that they are very low frequency devices and, as you know from the A-weighting curve, sound at low frequencies is not often much of a problem.

If you cannot personally supervise and adjust the installation, you can change the character of panel absorbers from rifles that sharpshoot the tuned frequency to shotguns that scatter their effect over a frequency range.

As you might expect, the shotgun panel is never as effective as the rifle panel is *at its best* frequency. Still, you don't have to aim as carefully!

The *"rifle" panel* is the same as the *"shotgun" panel,* except that there is a porous absorber behind the latter.

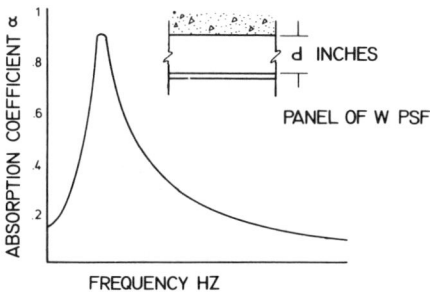

Figure 30 The simple panel absorber may be very sharply tuned. It has excellent performance at one frequency only—it is like a rifle. If this is the characteristic you want, you will have to aim it carefully by tuning to the frequency you want to absorb.

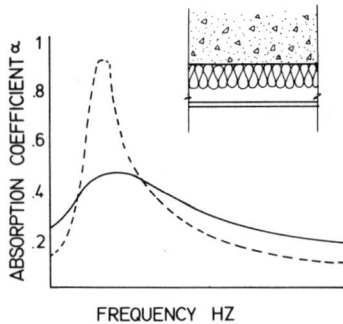

Figure 31 A modification of the panel absorber is often used. A layer of porous absorber in the cavity converts the rifle into a shotgun. Although not quite as effective at its best frequency as the simple panel absorber, this version does not have critical tuning. Performance shown solid (dashed curve is Figure 30 for comparison).

You will not find much supporting data to help you design panel absorbers. The best "test results" available are contained in articles about balancing the absorption in recording studios—and these are few and hard to find. The *idea* of panel absorbers is helpful, though. Does it explain to you the shape of the two following curves?

Figure 32 shows a curve (dashed) that peaks at 500 Hz. It looks much like the curve you might expect from a panel of the "rifle" type (with no absorber behind it). In fact, it is the performance of ⅛ in. pegboard with ⅛ in. holes on 1 in. centers over a 1 in. sheet of 3 pcf glass fiber board in a number 4 mount. The solid curve shows the performance of the glass fiber board alone. This is not a panel absorber because of the holes in the pegboard. It certainly isn't a "rifle" panel with the entire cavity filled with fuzz. It is a sharply tuned absorber patterned after the classic Helmholtz resonator.

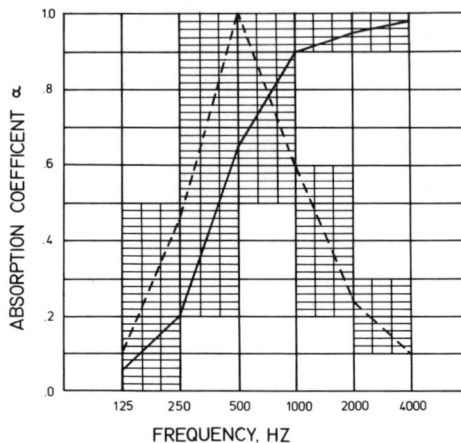

Figure 32 The solid curve is 1 in. glass fiber board in a number 4 mounting. The dashed curve is the same material and mounting but with a covering of ⅛ in. pegboard (⅛ in. Masonite with ⅛ in. holes on 1 in. centers). Do not jump to the conclusion that this is another panel absorber. (Compare the dashed curve here with the solid curve of Figure 30.) The perforated cover has formed a Helmholtz resonator.

HELMHOLTZ RESONATORS

Probably sometime you have used a soft drink bottle (or a jug) as a whistle. Everybody has done it. You supplied a source of power (the steady stream of air) and the bottle did the rest. The peculiar combination of volume of the bottle, the length of its neck, and the size of its mouth determine the tone you produced. The bottle is a tuned resonator.

Resonators—without their source of power—are also tuned absorbers. There is no simple *useful* formula that you need know but there are two times when you may find Helmholtz resonators helpful. One of these has to do with mufflers and is discussed in Chapter 10. The other is the use of slotted cinderblock and concrete block as a building material for walls.

When a wall of ordinary cinderblock has been laid up, it has an enclosed volume (the hollow cores) like a bottle. If it also had a mouth and neck, it could be a Helmholtz resonator. This has been done and the result is quite useful. Two characteristics of walls made of these slotted block may recommend them to you in some instances.

The first is an acoustical characteristic: it happens that for the enclosed volume (in typical wall construction) and the arrangement of slots, these blocks have very good low frequency absorption as an inherent property. There are also modified designs that offer a wider frequency range and so forth, but the inherent low frequency performance ought to interest you. You have just seen how hard it is to absorb lows.

The second characteristic is an economic one. Let us say you are working for a publishing company and a new printing plant is going to be built. Being an old printer yourself, you know that low frequency noise dominates the press room. In building a new plant, you might suggest going one of two ways:

1 Build a good conventional building with *no frills.* It saves some money and the noise control can come later (at about $12 to $18 per square foot of treated wall space—Consultant's note).
2 Since you have to build the wall out of something, use slotted hollow block. Yes, it will cost $1 to $2 per square foot premium (it's a tough one to sell.—Consultant's note).

(Obviously you will have mustered your facts and called for only the amount of premium wall that is reasonably supported by the predictable noise problems. The idea didn't sell for one consultant who will have to take on the worse job of straightening out "noise control" that "can come later" for a big metropolitan city's leading newspaper.)

WRAPPING UP

Before you begin to solve real problems with absorption, you should consider the following helpful suggestions.

First, you must understand that what has been presented here under the topic of absorption is a pretty broadbrush approach. Do not hesitate to work out problems on the basis of what you have learned—the journeyman acoustician does most of his or her work with absorption on no more sophisticated basis than you have before you. However, he or she knows when some special situation means that the whole process is risky and another method ought to be looked at—or when some simple cause leads to the noise problem and no heroic number of square feet of absorption need to be considered.

The best advice you could have, when you have a situation where absorption looks like the answer, is to figure out what the cost will be. If it is not too high to get you in trouble, try it. Don't cut your teeth on a job that is going to run to six figures. Some of them do.

The most difficult part of your calculation/data-gathering task is to make an inventory of the absorption originally present in a large room. Learn to estimate distances and dimensions (you can check them out with the blueprint later) and pay attention to the surfaces that are more absorptive (open doors, open roof top ventilators, ventilation duct, etc., are easy to miss but are *big* items in your inventory). Use Table 5 for values of alpha and any other reliable data you can find. Then, begin to use your experience—as you gain it—to estimate what the alpha will be for some new material that is like something you already know.

Table 5 omits several materials with which you will frequently be working. You will not find steel, for example, although you will encounter it on almost every job. It cannot be handled easily because a steel I-beam is utterly different from a 24-gauge steel ventilating duct and neither of them is like a steel door $1^3/4$ in. thick and packed with mineral wool. When you face the problem of working with such a door, for example, look through the table and see what similar constructions or materials you can find. If the surface in question is a small part of the total absorption, you can make a plausible guess about its behavior from other data or even lump it into a miscellaneous term. If it is a large part of the total absorption, you're in trouble!

Don't stop with the room surfaces or ductwork running through it. (As a ridiculous extreme, consider that glass fiber board is made in a room, too!) Look around at *everything* that absorbs sound.

Allow for *excess* absorption in a room with big panels, lots of cavities in and under equipment, soft materials like sawdust or lightweight trash (paper, fabric, chips, etc.), even the people themselves if they're wearing heavy clothes. Excess absorption is guesswork to some extent, and excess absorption can be overdone. Still, it keeps you on the conservative side when you are estimating that critical "A_s."

If you do overdo the excess absorption or are too conservative in estimating the original absorption, you will call for too much added absorption—and too much money.

This is exactly the problem already mentioned—estimating absorption for large surfaces that might be very important but where no good data are available.

An excellent way to avoid this is to install the new absorption in stages (different materials or surfaces, perhaps) and *measure the noise reduction achieved in the intermediate stages.* This will let you figure back to what the original absorption really was. You can then be a hero with the announcement, "Guess what, boss, we can save $20,000 on the A-wing project."

THE SURE WAY

With the right instrumentation available, you need not estimate the original absorption at all. In this case, you measure the "reverberation time" of the space. This is the time it takes for sound of a specified frequency (actually a band of frequencies—typically an octave) to decay 60 dB. In other words, if the sound source suddenly stops, the time it takes for the sound to disappear is a function of the amount of absorption that is taking it out of circulation. The simple equation is

$$T_{60} = \frac{0.05 \ V}{A}$$

where T_{60} = time in seconds for the sound pressure level of the specified band to fall 60 dB
V = room volume (ft^3)
A = absorption (sabins)

There are more sophisticated equations for this, like the Norris-Eyring equation. Account may also be taken of absorption by the air itself in large rooms. By the time you have the measuring equipment to use these approaches, you will know all about them. They are rarely required to get practical results.

GENERAL GUIDELINES

In appraising the chance of success for absorption to control noise, the following circumstances argue in its favor:

1 The room is finished with "hard" (nonabsorptive) surfaces.
2 There are no conspicuous absorbers present as large openings (as a fraction of the total room surface) or as materials in process or storage.
3 The ceiling is low and the area large (or, rarely, the room is long and tunnel-like)
4 The requirement is to protect people far removed from the sources of noise.
5 Focusing (curved or corner) surfaces are directing sound to people

Be skeptical whenever you find locations near noise sources are a problem. And beware of the salesman who says it takes care of everything.

SOMETIMES ABSORBERS WORK NEAR THE SOURCE

Sometimes absorption *does* work near the source. You need to make a nice distinction between the two cases of (1) reducing reverberant noise fields (the usual job assigned to absorption) and (2) prevention of noise reflection from surfaces near the source. In looking at the free field (or direct field), we looked at Q. Another way of picturing what Q means is given in the following table:

Location of the Source	Q	Increase in Sound Pressure Level (dB)	Fraction of Energy Reflected from the Nearby Surface(s)
In free space	1	0	0
Near a large flat	2	3	0.5
In a simple corner	4	6	0.75
In a trihedral corner	8	9	0.875

By covering the nearby reflecting surface(s) with the right absorber, you can do away with the *reflected fraction* of the sound energy. You can look at the potential gain more directly in the center column of the table. When the noise source is quite close to reflecting surfaces, the use of absorption is more efficient than when the source is far from them. Consider that the energy impinging on surfaces near the source is much more concentrated—more energy per square foot—than it is farther away. Thus a little absorption on the surfaces near the source can be as effective as much absorption farther away in the room.

The reason for the repeated caution about using absorption for locations near the source is that the absorption can never have an effect on the noise until the waves are reflected at least once. The direct field of the source will continue to be substantial.

In the usual case where people are closer to the source than the reflecting surfaces are, small (sometimes unmeasurable) gains are all that is possible by covering the reflecting surface. The benefits of using absorption in cases like this are only realized away from the source.

PROBLEMS

1 A worked problem emphasizing (1) estimation of noise reduction by using absorption, (2) checking cost effectiveness, and (3) being sure to consider the "nonacoustical" aspects of the problem. A new master alloy room is to be built. It will be identical in dimensions and finish to one in operation at another plant. It will also be equipped in the same way with two small gas-fired reverberatory furnaces. Since the existing room has a level of 98 dBA with both furnaces in operation, there is some interest in using hollow, slotted cinderblock to replace some of the ordinary block in building the walls. The contractor estimates the cost premium for the slotted block will be about \$1.00/ft², in place. Alternatively, the walls could be built with conventional block and hanging functional absorbers installed. The cost estimate for the absorbers, in place, is \$2.50/ft² (based on one side—i.e., a 2 × 4 ft absorber is considered to be 8 ft²). Which treatment of the room will produce the lower noise level for the same cost?

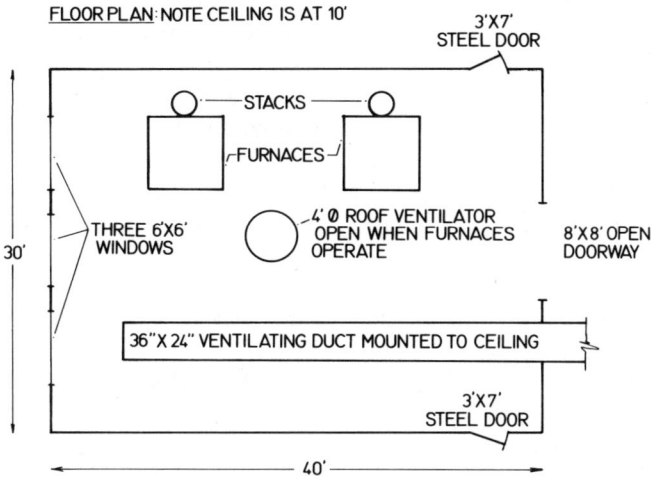

FLOOR PLAN: NOTE CEILING IS AT 10'

3'X7' STEEL DOOR

STACKS

FURNACES

4' Ø ROOF VENTILATOR OPEN WHEN FURNACES OPERATE

THREE 6'X6' WINDOWS

8'X8' OPEN DOORWAY

30'

36"X 24" VENTILATING DUCT MOUNTED TO CEILING

3'X7' STEEL DOOR

40'

Data

| Spectrum in Existing Room with Both Furnaces Running | | | | Absorption Coefficients | | |
| | | | | | 2 × 4 ft × 1 in. Hanging Absorbers | |
Frequency Octave Band (Hz)	L_p (dB)	A-weight Correla- tion	L_{eff}	Slotted Block	Type 1	Type 2
63	90	−26	64	—	—	—
125	101	−16	85*	0.61	0.05	0.05
250	102	−9	93*	0.91	0.46	0.28
500	96	−3	93*	0.65	0.92	0.58
1000	93	0	93*	0.65	0.83	0.71
2000	79	+1	80	0.42	0.58	0.88
4000	70	+1	71	0.49	0.27	0.93
8000	59	−1	58	—	—	—

*Note that the A-weighted level will be controlled by these four bands alone.

Room finish (see preceding sketch, ceiling height is 10 ft):

Item	Area (ft^2)	Material
Windows	108	Single glazed, ⅛ in. glass
Doors	42	Steel, mineral wool filled
Open doorway	64	Open to quiet area
Balance of wall	1186	8 in. cinderblock, painted (note some may be replaced with slotted block)
Floor	1200	Painted concrete
Ceiling*	1091	Painted concrete
Roof ventilator	13	Open to sky
Ventilating duct†	216	36 × 24 in. × 30 ft of 24 gauge galvanized
Miscellaneous		
Two stacks	126	Bare brick
Two people		
All other		Allow 30 sabins

*1200 ft^2 less ventilator duct and stacks.
†Three exposed sides and end only. Remaining side is flush with ceiling.

Method

(a)   There are many ways you might get started on this problem. One obvious question you will have is: How much noise reduction do we need to achieve? If you need only 1 to 3 dBA, either method (blocks or hanging absorbers) may work. If you need much more, there is a question of whether there will be enough space for the hanging absorbers. Notice that there is only 10 ft of headroom and there are parts of the ceiling that cannot be covered by hanging absorbers (stack, roof ventilators, space over the furnaces). No overhead crane was mentioned, either. If an industrial truck is used for loading and heavy lifting, you will not be able to hang absorbers under the duct for lack of headroom.

(b)   Based on these practical considerations, sketch a layout to find how many functional absorbers will fit in the available space. They are sold in packages of 24. Do not space them so closely that they starve each other. You will find that 48 absorbers is a reasonable number to use. This tells you that if you need a substantial noise reduction in the room, functional absorbers cannot do it in this case.

(c)   To compare effectiveness of the two methods, find the noise reduction the 48 functional absorbers will produce. The cost will be $960 (48 × 2 ft × 4 ft × $2.50/ft^2). See what noise reduction is available from 960 ft^2 of slotted block.

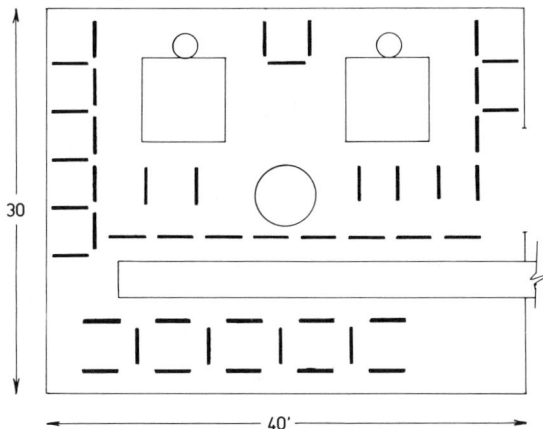

Calculations

Find absorption in existing room as the reference.

Item	Area (ft^2)	Absorption Coefficient (for critical bands) (Hz) Absorption, sabins, for critical bands			
		125	250	500	1000
Windows	108	0.03	0.03	0.03	0.03
		3	3	3	3
Doors	42	0.1	0.06	0.03	—
		4	3	1	0
Open doorway	64	64	64	64	64
Balance of wall	1186	0.10	0.05	0.06	0.07
		119	59	71	83
Floor	1200	0.01	0.01	0.01	0.02
		12	12	12	24
Ceiling	1091	0.01	0.01	0.01	0.02
		11	11	11	22
Ventilator	13	13	13	13	13
Duct	216	0.15	0.08	0.05	0.02
		32	17	11	4
Stacks	126	0.02	0.02	0.03	0.04
		3	3	4	5
People (2)		5	7	8	9
Other		30	30	30	30
Total sabins		296	222	228	257

Absorption in this room (or the new room if conventional block is used) will be the total shown above for the four critical bands. If 48 of the hanging functional absorbers* are added they will cover none of the absorption surface just evaluated and the absorption present will increase as shown by the following data:

*Type 1 absorbers were chosen rather than Type 2. Why? Which type would be better if the operation were sandblasting with a peak in the spectrum at 2000 or 4000 Hz?

Item	Area (ft^2)	Absorption Coefficient (for critical bands) Absorption, sabins, for critical bands			
		125	250	500	1000
48 (Type 1) absorbers	384	**0.05**	**0.46**	**0.92**	**0.83**
		19	119	353	319
New total, including absorbers		315	341	581	576
Noise reduction = 10 log (A_f/A_s)		0	2	4	4 dB
Existing room level L_{eff}		85	93	93	93 (98 dBA)
After installing absorbers, L_{eff}		85	91	89	89

—or to 95 dBA. Thus a 3 dBA noise reduction was achieved at a cost of $960. That cost is now applied to new construction using slotted block so that its effect may be estimated and compared.

Item	Area (ft^2)	Absorption Coefficient Absorption, sabins, for critical bands			
Window		3	3	3	3
Doors		4	3	1	0
Doorway		64	64	64	64
Slotted block	960	**0.61**	**0.91**	**0.65**	**0.65**
		586	874	624	624
Balance of wall	226	**0.10**	**0.05**	**0.06**	**0.07**
		23	11	14	16
Floor		12	12	12	24
Ceiling		11	11	11	22
Ventilator		13	13	13	13
Duct		32	17	11	4
Stacks		3	3	4	5
People		5	7	8	9
Other		30	30	30	30

Total sabins, with blocks	786	1048	795	814
Total sabins, existing room	296	222	228	257
Noise reduction = 10 log (A_f/A_s)	4	7	5	5 dB
Existing room level L_{eff}	85	93	93	93 or 98 dBA
After adding blocks L_{eff}	81	86	88	88

—or to 93 dBA, so the block accomplished a 5 dBA noise reduction for the same $960 cost.

There is a moral to this problem. It is *not* that you should always use slotted block. The new construction advantage and the shape of the spectrum both favor the block in this case.

8

SINGLE-WALL
TRANSMISSION LOSS

Several noise control methods depend on putting an obstacle in an air path that sound takes. The effectiveness of such a noise control measure depends on:

1 How much sound can get around or through holes in the obstacle.
2 How much sound can get "through" the obstacle itself in the sense that the sound makes it vibrate and thus reradiate sound on the far, or quiet, side.

This chapter and Chapter 9 deal with the general idea. In both chapters and in noise control in general there are some frequently used definitions. The first three of these give more precision to the meaning of *obstacle.*

A *wall* is a surface sealed on all edges so that there is no important way sound on one side of it can get to the other except by going "through" it (i.e., making it shake and reradiate). Real walls often have doorways, cracks at the periphery, or back-to-back electrical boxes, or a host of other things including simple holes. To make good sense of noise control by using walls you will need to find what a perfect, unblemished wall of the sort you propose would do *and then* find out how that situation will be altered by all the ways sound can leak around (or *flank*) it.

A *barrier* (sometimes indexed as a "partial barrier") is an obstacle (often wall-like but it also may be a building or any object) that sound may go around on one or more edges as well as going through.

An *enclosure* is made of walls (including the ceiling and floor) and like walls must first be considered in an ideal way and then reexamined to allow for the effects of openings and so on.

In this chapter you can ignore all the ways in which sound can get around your wall. This chapter is concerned solely with how much sound can get through a wall by shaking it to make it reradiate on the far side. Moreover, this chapter is only concerned with single walls.

Single walls are walls that consist basically of a simple sheet of material. The most complicated single wall might have framing to hold it up and its sheet might be several materials joined together (stucco on brick, shingles on sheathing over studs, or a carpet on a finished floor supported by a rough floor supported by joists) but a single wall has no big cavities in it. Chances are that as you read this any wall you can see has studs in it. They trap big air cavities and make that wall a double wall. These are discussed in Chapter 9.

TRANSMISSION LOSS

When you look for how much sound can get through a wall by the shake-and-reradiate path you are looking for a quantity called *transmission loss* (TL). Transmission loss is the standard figure of merit for all walls. There are two important things you should know about TL.

First, although TL is given in decibels, it is used with rational arithmetic rather than the convoluted routine with which you've learned to treat decibels. For example, if a certain location has an L_p of 100 dB and you put a wall with a TL of 40 dB between the only source of sound and that location you can expect the level to drop to 60 dB. (From the previous chapters on absorption you've learned that such a statement can only be true if the source produced a single tone and the rating of the wall was for the frequency of that tone. That's quibbling!) The idea is that you subtract the decibels of TL as if they were apples. No esoteric math is involved with using TL.

Second, the truth of the last paragraph *usually* applies. Actually, TL is an idealized concept and in some cases you do have to do a little massaging of the data to find the noise reduction (which is the real payoff figure). Remember that TL is about the right answer in most cases. The difference between NR and TL is given as

$$NR = TL + 10 \log (A/S)$$

where   NR = noise reduction (dB)
         TL = transmission loss (dB)
          A = absorption in the receiving room (sabins)
          S = area of the wall (ft^2)

In run-of-the-mill situations in the real world, either (a) A and S are close enough that the effect of 10 log (A/S) is minor or (b) A is big compared to S and the effect of the wall or enclosure is much better than you would expect if you didn't allow for the difference between TL and NR. You need to be alert to the difference when the receiving room is small and hard. Then the difference between NR and TL can be appreciable if the wall surface involved is large. Try to think of a real situation that meets those requirements. A small hard (reverberant) room where you need to protect against an outside source by choosing four doors would be such a situation. They're not common.

WHY ESTIMATE TL?

There is an abundance of TL data for common structures—perhaps even more data than for absorption. Why, then, bother to learn to estimate TL? There are two excellent reasons.

First, as with absorptive materials, the biggest market is for general architectural use and it follows that the bulk of the data are for gypsum board, plaster, wooden doors, and so forth. Industrial noise control makes other demands, for example: immunity to water or corrosives, high temperatures, and extreme strength requirements.

Do you suppose you could look up the TL of ¼ in. steel plate in published data?

As a matter of fact, you could—though it requires having *a lot* of data and not the sort that manufacturers hand out for free. And of the two tests you might find, one is woefully wrong even though it was made (long ago!) at the National Bureau of Standards. Suppose you accepted such excellent credentials as those of the Bureau of Standards and based an expensive job on those data!

Estimating the TL would give you a clue that something was the matter. After rechecking your calculations for your error a few times (and finding none) you would at least have a warning. As a matter of fact, you would have a fairly reliable estimate of the performance—much better than the published curve. And the calculation is fairly simple. It comes up shortly.

Remember the second reason: Not all published data are reliable! In the first place, techniques of measuring and testing TL have improved. If you can find the test date and it is 10 years old, a check calculation is a good idea. In the second place, some test labs consistently find results different from the average results of all labs. In the third place, by testing the "same" material several times in different labs and with slightly different detailing of construction, it is possible to get a range of at *least* a few decibels. Now if you were a reasonably honest material manufacturer with a healthy competive spirit, which test would you publish?

CAN YOU BELIEVE YOUR ESTIMATE?

The theory behind TL gets tricky and you will want some experience behind you before you project the transmission loss of a complicated structure. This is not a problem with simple structures of ordinary size, incidentally, even though the materials may be a little out of the ordinary.

The estimation methods given here are like mongrel pups. Their pedigree is not their virtue. Their virtue is that they work. By all means read every article you see on TL. Some of the theories and estimation methods you see will have very good pedigrees indeed—yet many of the estimation methods based on theory alone can get you in serious trouble.

A common failing of *all* methods of estimating TL is that in order to test them, clean data are required. *Clean data on actual field installations are a contradiction in terms!* The most common complaint about TL is "But you said . . . but the data said . . . but the test said!" To establish performance, use of lab test data of TL is the only sane basis. The ability to predict what is going to happen in a well-controlled lab test is not magical. What is installed in the field, however, is frequently not what was tested in the lab.

One common wall is well-known to have a TL rating [actually, an STC (sound transmission class), which will be discussed in Chapter 9] of 37. Think of that number as a TL rating in decibels. To determine improvements that could be made by remedial work *after the wall as typically erected by union carpenters* was already in place, the wall was put up without supervision by union carpenters in a highly respected test lab. It tested at 29!

Moral Whether you estimate the TL performance or rely on test data *test what you are going to build* (warts and all!) *and build what you tested* (or estimated).

MASS LAW

If the wall you have in mind is quite large and absolutely homogeneous you will need only mass law and one other effect to predict how it will behave. Mass law says that the TL of the wall will rise at 6 dB per octave and will have a TL at 500 Hz of

$$TL_{500} = 20 \log W + 21 \text{ (dB)}$$

where TL_{500} = transmission loss at 500 Hz (dB)
    W = the wall surface weight (psf)

This is a nice, smooth curve. The data at 500 Hz, from lab tests, confirm that it gives you a fairly accurate value. The slope is open to a little more question. *Six* dB per octave is the right theoretical slope. A better opinion, after looking at many test data, is that 5.5 dB is closer, for the real world. Why is 6 dB the theoretical slope? In the *mass* law region of the curve, mass alone controls the TL. Nothing in physical science ever gets more than a stone's throw away from Newton's

$$F = ma$$

We account for the mass by looking at the surface weight of the wall in pounds per square foot. (True, weight isn't mass, but as long as you are building at sea level, on earth, mass and weight are related by a constant—and if you are finding mass law behavior on the moon, you're on your own.) The same square foot will convert the pressure to a force, which accounts for F and m.

Suppose now that we assume a constant sound pressure (dynamic pressure) on one side of that wall at all frequencies. F is constant; m is constant. Therefore, acceleration is constant. What is not constant with varying fre-

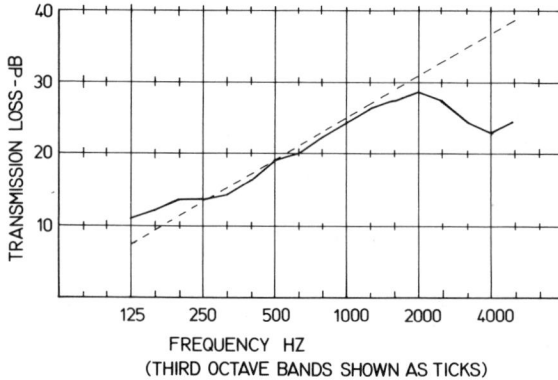

Figure 33 Transmission loss of ¼ in. plywood from test data. The dashed line is the theoretical 6 dB/octave slope. Although agreement at 500 Hz is good, a slightly lower slope would have produced a better match for the "straight" portion of the curve.

quency is the time in each cycle of sound pressure variation for the wall to deflect.

It is the total deflection on the far side of the wall that will recreate the sound wave there. How is deflection related to acceleration when the frequency changes by one octave? The time available is cut in half. The acceleration is constant. Therefore, the displacement is $(\frac{1}{2})^2$ or one quarter as great.

Now this is acoustics, so you just know that we are about to take a log. If you like we can take the log of ¼ and the log is *minus* 0.6 (for those of you with calculators) and $-1 + 0.4$ if you come from ancient times (same answer, of course). Or you could just say the ratio is 4 and the log 0.6, knowing all along whether to use a plus or minus sign in front, depending on which way you are going in frequency.

At any rate, from this discussion it should be clear that for a simple, single wall controlled by mass, less and less sound is going to get "through" the wall as frequency becomes higher. We have already established that each octave change in frequency lowers the sound getting through by 6 dB ($=10$ log ratio of displacement). If the sound getting through is 6 dB less, the TL is 6 dB more.

SO MUCH FOR FREQUENCY

If you double the mass of the wall, a similar effect takes place. If you look up the underlying theory, you'll find that the sound pressure on the quiet side of the wall is inversely proportional to the wall's mass. But the sound power is proportional to the *square* of that pressure. Doubling the mass therefore quarters the power. The factor is 4, the log 0.6, and the change in TL 6 dB.

Intuition, again, must tell you that a heavier wall will stop more sound from

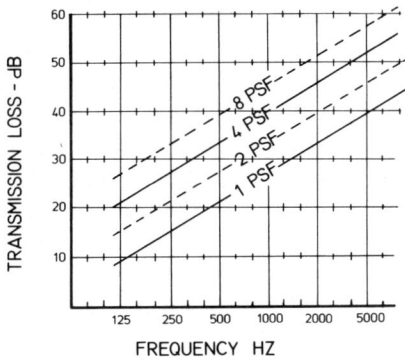

Figure 34 Every time the mass of the wall is doubled, the mass law TL increases 6 dB. The theory is reliable and the results are well borne out in practice.

getting through than a light one, so you know where to put the plus or minus sign. Doubling the weight of a wall improves its *mass law performance* 6 dB.

Believe it or not, you have just covered what takes pages of equations in theoretical derivations: 6 dB for doubling of frequency, 6 dB for doubling of mass. Is it true? It's good enough not to get you in trouble for most practical structures.

MASS LAW IN A NUTSHELL

If mass law were the only thing to worry about you could wrap it all up in a single equation:

$$TL = 20 \log W + 18 \log F - 28$$

where W = surface weight (psf)
F = frequency (Hz)
TL = transmission loss (dB)

The slight difference between the two coefficients 20 and 18 in the preceding equation reflect the fact that doubling the mass of the wall really does produce a 6 dB increase in performance in the mass law region. Doubling the frequency does not usually produce quite a 6 dB increase in TL in this region. The coefficient of 18 will give the curve a slope of 5.5 dB per octave, a reasonable average of the performance for a wide variety of materials *and test sites*.

A COMPLICATED COINCIDENCE

For a single wall your ability to predict performance by the mass law alone ends at a frequency where another effect—the "coincidence effect"—takes over.

At some time you have shaken out a carpet, so you are thoroughly familiar with the idea of a bending wave in a sheet. If you can visualize what is happening in the room you are in now, or in some noisy room in your plant, it will be apparent that direct and reflected sound waves are traveling in many directions. Think of the waves that approach a wall of the room from a fairly low angle, say 30°, so that they are grazing the wall surface. At every pressure front these waves must be pushing the wall in; at every rarefaction front, "pulling" it out (or at least the air pressure on the far side is pushing it into the room). These inward and outward bulges in the wall tend to move along with the sound wave.

We have already decided that when the far side of the wall is made to move by a sound wave, part of the sound has gone "through" the wall. Now these bending waves that are forced into a single wall by sound waves may seem unremarkable. For the most part, they are; they are no more degrading to the wall's TL than sound waves that hit the wall head on (at right angles) and force it to move like a piston.

However, an interesting fact about bending waves in sheet material causes serious loss of TL in walls at some frequencies. The velocity of bending waves in sheets is comparable to the speed of sound in air. *And* the velocity of the bending waves increases with their frequency. This means that the wavelength of a sound wave in air that approaches the wall at an angle may project a bending wave into the sheet *at exactly the velocity the bending wave likes to travel.* The projected sound wave and the bending wave *coincide*—hence, "coincidence."

Now for a given sheet of material, there is no single frequency at which you can expect coincidence. Rather, as shown in Figure 35, there is a range of frequencies, and their matching angles, at which the two waves coincide. Over part of this range, the match between the two waves is so good that between 10 and 100 times as much power is coupled into the wall as you would expect from mass law prediction. This translates into 10 to 20 dB more sound going "through" the wall—typical of what you will find in the "coincidence dip."

Stop for a moment and review what you have learned! Poor performance in the coincidence range is caused by bending waves in the sheet. The range of frequencies over which the coincidence effect is important is governed by the angles at which the sound arrives at the wall. If these two points are clear, you are well on the way to a useful understanding of coincidence.

For example, would you expect to find severe coincidence effects in a small sheet of material? No, because the dimensions of the sheet need to be big enough to allow bending waves to develop. Will conventional tests (9 × 14 ft walls tested in large reverberant rooms) tell you much about the coincidence dip to be expected when the noise is projected at the wall from a discrete angle? No, because a *reverberation* room implies that the sound waves arrived from all angles, randomly.

DAMPING

Chapter 12 is devoted to the subject of damping. A peek ahead now is required in order to make sense of coincidence. (As a matter of fact the difference be-

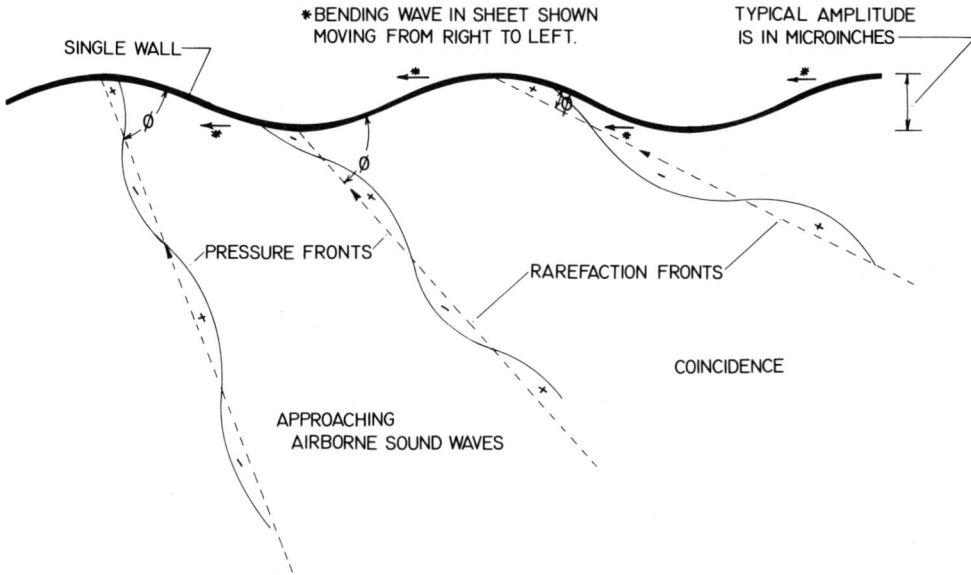

SINGLE WALL

✱BENDING WAVE IN SHEET SHOWN
MOVING FROM RIGHT TO LEFT.

TYPICAL AMPLITUDE
IS IN MICROINCHES

PRESSURE FRONTS

RAREFACTION FRONTS

COINCIDENCE

APPROACHING
AIRBORNE SOUND WAVES

Figure 35 Airborne noise of various frequencies (wavelengths) and angles of incidence ϕ with bending waves in the sheet. For clarity, the bending waves are shown moving in one direction. In a real wall there would be bending waves moving left, right, up and down, and in all other directions.

tween the "rifle" and "shotgun" panel absorbers covered in the last chapter is a special case of damping in action also.)

In a nutshell, damping is that mechanism in a resonant system that drains energy out of the system. This can happen in so many ways that in many common systems several mechanisms damp the oscillations.

One such common system is the very wall we are interested in here. First, remember that we have been talking about a simple, single, homogeneous sheet of material. Ideally, if all the bending wave activity is going to work as theory suggests, the sheet ought to be infinite in extent. Aluminum, steel, and window glass are good examples of simple, single, homogeneous sheets. But infinite extent?

Even where the sheet is so large that it does not cramp the theory of how coincidence ought to effect the TL, it will usually have some sort of framing members. Beyond that, the framing members are physically attached to the system by means which themselves are often not simple—a steel sheet welded to a steel frame member is different from a steel sheet riveted to a steel member, a sheet of gypsum nailed to a stud, a sheet of aluminum glued to a frame, a sheet of glass puttied into a wood frame or supported in a rubber boot in a steel frame, and so on.

To the extent that the sheet material is *not* homogenous (brick *and* mortar), not infinite in extent (!), or is fastened *somehow* to *something* (not simple or single), the whole mechanics of bending waves can only be an idealization of the coincidence effect. Energy will be lost because the real wall has places where it loses energy to its own discontinuities, to mismatched materials at its edge, or to imperfectly matched parts like framing members and their fastenings. These places where energy is drained from the system (turned into heat by friction in some sense) constitute the damping of the system.

The TL curve dips in the coincidence region because the wall likes to bend at the frequencies in that region. Damping limits the amount of bending—and hence the depth of the dip—by using the energy that makes the wall vibrate. Some of the places where this may happen are at fastenings and sheet/frame joints by sliding friction; at the putty or rubber boot around a pane of glass by nonelastic deformation; or within the material itself if it is not continuous (brick and mortar) or homogeneous (lead-loaded vinyl).

As might be expected, damping also reduces the frequency range over which the coincidence effect is important. In construction, where a material like sheet steel is fastened in place with minimum damping, the coincidence dip will extend over almost four octaves. A well damped material like leaded vinyl shows a dip barely more than two octaves wide.

STIFFNESS

Since the key to coincidence is bending waves in the sheet, it is no surprise that stiffness is the key to coincidence behavior. As sheets become stiffer the frequency at which the dip starts drops lower and lower. This is evident in Figure 37 showing the TL for several materials when they are used in sheets weighing 1 psf (pound per square foot). A 1 psf sheet of lead is only 17 mils thick. It is quite limp. A 1 psf sheet of plywood is about 5/16 in. thick and is harder to flex than any of the other materials shown.

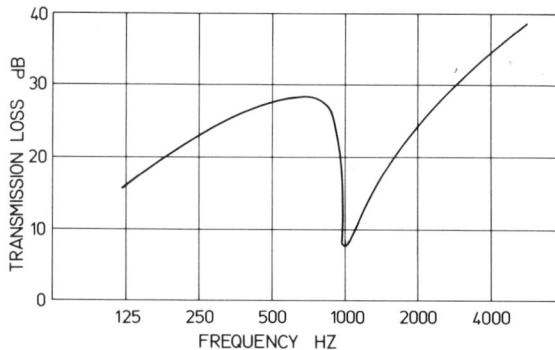

Figure 36 You are unlikely to ever see coincidence behavior this severe. It is what you might expect of a system with so little damping that only 1% of the energy is being drained out.

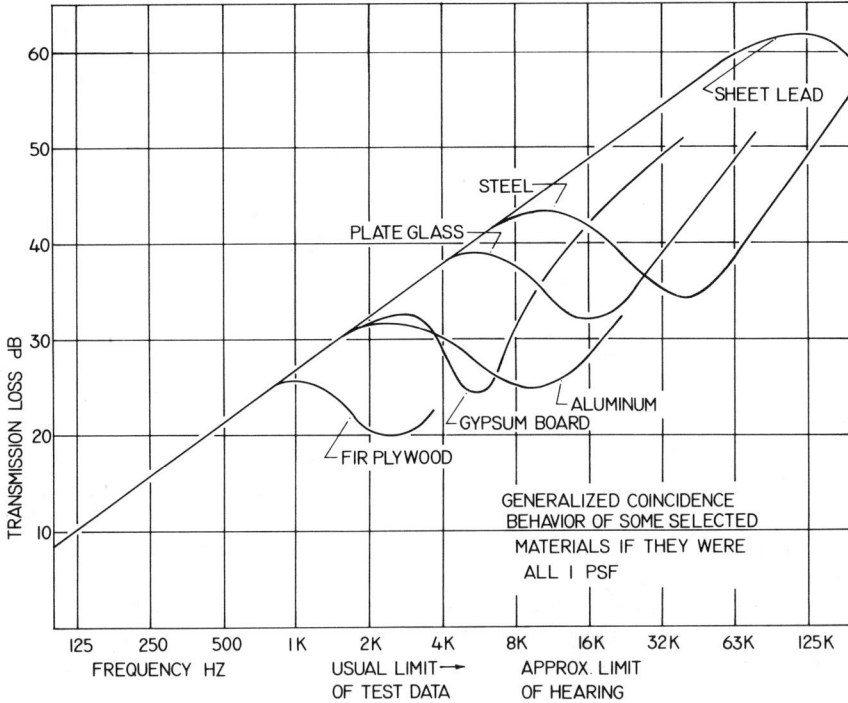

Figure 37 Predicted coincidence behavior for sheets of material weighing 1 psf. Plywood of this weight is about ⁵/16 in. thick and lead is 0.017 in. thick. Of all the sheets, plywood is the stiffest and goes into coincidence first.

These design curves are used to predict TL performance of structures. They work for walls of usual dimensions (bigger than a doghouse is a good rule of thumb—certainly comparable to the wavelength of the sound in air).

Consider, now, a sheet of ⁵/16 in. plywood that falls within the size requirement. Its TL curve shows a departure from the mass law line at about 630 Hz. This is the beginning of the coincidence dip. What would you expect for TL from a sheet of ⁵/8 in. plywood? Mass law tells you that the TL should be 6 dB better because ⁵/8 in. ply weighs about 2 psf instead of 1. But doubling the thickness also increased the stiffness—and very considerably.

Mother Nature gave noise control people a real break here. She arranged matters so that the increase in stiffness when the thickness of a sheet is doubled makes the beginning of coincidence fall just one octave lower. The ⁵/8 in. plywood will go into coincidence at 315 Hz. Now that may not seem like much of a break until you compare the two curves in Figure 38. Doubling the weight caused a 6 dB increase in TL and a one octave shift in coincidence *along a line with a 6 dB per octave slope.* Coincidence for the plywood sheet started at 23 dB of TL in both cases!

In fact, 23 dB is a characteristic TL for plywood of any thickness. The sheet

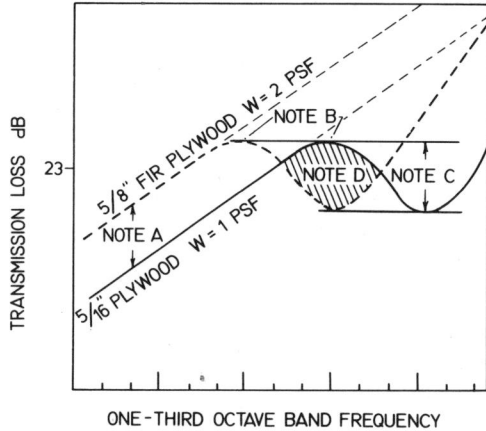

Figure 38 A close-up of the coincidence region for two sheets of plywood ⁵⁄₁₆ and ⁵⁄₈ in. thick. Note (*A*) that the thicker sheet weighs twice as much as the thinner and its mass law performance has increased 6 dB over the lighter one. Note (*B*) that the difference in stiffness between the sheets brings the heavier one into coincidence about one octave before (lower frequency) the light one. This means, Note (*C*), that the upper and lower excursions of the curve stay within the same limits for plywood (or any other material) in the coincidence region. Finally, Note (*D*), there is a range of frequency where the heavier sheet does not perform as well as the light one.

will always go into coincidence at that TL. There is a characteristic figure for every material called the plateau height. The plateau height for plywood is often given as 19 though 23 has always worked better in the author's experience. For aluminum it is 29, for plate glass 33, and so forth.

THE SINGLE WALL TL CURVE

In estimating the performance of a single wall, you always start with the mass law. When the curve reaches the plateau height it departs from the mass law line. In some cases it really does fall in a fairly level plateau, but this is the exception and not the rule. It is better to assume that the curve will undulate over a range of frequency called the plateau breadth. (These two figures for the plateau are given in Table 7 for several common materials.)

In the first quarter (or somewhat more) of the plateau breadth, the TL will rise only 3 dB from the characteristic plateau height. In the last quarter (or somewhat less) of the plateau breadth it will drop to about 4 or 5 dB below plateau height. Gypsum board is unusual in having a dip about 7 dB below plateau height in the typical sizes of sheet in normal construction. Finally the curve will climb and will have a slope of 10 dB per octave as it leaves the coincidence region.

Table 7   *Generalized Coincidence Data*

Material	Specific Surface Weight (psf at 1 in. thick)	Plateau Height (dB)	Plateau Breadth (one-third octave bands)
Aluminum	14.2	29	11[7]
Brick	11	37	8
Chipboard[1]	3.3	34	8
Cinderblock			
Low density	5	33	9
High density	8	33	9
Concrete (typical)	12	30	10
Deciban	1.6	30	8
Fiberglass[2]	9	~30	8
Glass (plate)	13	33	9
Gypsum board[3]			
Single sheet	4	31	7
Double sheet[4]	4	38	7
Hardboard[5]	5	34	9
Lead	59	56	7
Leaded vinyl	24	60	7
Plank (pine)	2.4 (typical)	20	8
Plaster[6]	9	30	9
Plexiglas (Lucite)	5.6	27	8
Plywood (fir)	3	23	8
Stainless steel	41.6	36	10[7]
Steel, mild	41.8	40	11[7]
Transite (millboard)	9	35	9

[1]Novaply, "flake."
[2]FRP, fiber reinforced plastic, as used in boats.
[3]Though there are minor variations among grades and manufacturers, these data are generally applicable.
[4]For typical mechanical fastening. If laminated with adhesive, use these data, but allow for some uncertainty. Plateau height *may* rise.
[5]Masonite.
[6]Average of lime and gypsum data. Applicable to both.
[7]Under worst conditions. Will be narrower in damping mounting.

Fastening and framing details and other factors may cause the curve to be different from that just described. For usual construction methods the method just given is a reliable guide to the performance you can expect.

Some exceptions worth looking at are:

1   Sheet glass is quite variable in its coincidence behavior. If it is bedded in putty or mastic or mounted in a lossy rubber or plastic boot, it can have an

almost flat plateau. On the other hand, when glass is used in large expanses and rigidly supported, it can have an unusually sharp and deep coincidence dip. The ground floor lobby of the Union Carbide Building in New York is walled in glass and the dimensions of the sheets are heroic. If the lobby is quiet, you will hear a narrow band of frequencies in the lobby—though the source of the noise, the traffic outside, is broadband.

2   Damped sheets will do better in coincidence than undamped sheets. Some common tricks for damping are bedding the sheet in calking at edges and framing members, using foam builders' tape for the same purpose, adhering sheet lead or leaded vinyl to the sheet, or coating it with a heavy mastic.

3   All bets are off when you work with corrugated sheets. The ridges or corrugations drastically stiffen the sheet in one dimension and, by acting like hinge lines, make it limper in the other. Such "sheet" materials may even have two coincidence dips.

Using mass law and the generalized information of Table 7, you can work out the TL curve for single walls of any of the materials listed. The table will furnish general guidance in other cases, too. Look for a material in the table that is similar to the one you are working with.

SPLITTING HAIRS

If you look up the actual test data from a good lab, you will seldom find the nice smooth coincidence region shown in Figures 37 and 38. Instead of being shaped like a sine wave, the curve is stretched out in the first half and crowded together in the second. If you are painstaking, you can make a better approximation of the region by dividing the frequency range into quarters numerically instead of graphically on the log paper. The peak and dip still fall at the first and third quarters.

The difference between the two methods is shown in Figure 39 for a sheet of gypsum board weighing 1.78 psf and tabulated in Table 9.

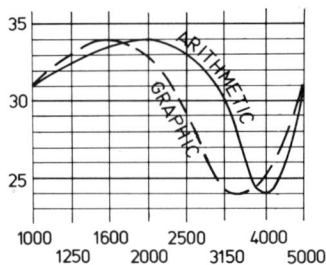

Figure 39   Compared to actual test data the shape of the curve in the coincidence region can be better approximated by using the "arithmetic" method. The precision of prediction in the coincidence region is not all that good and the simpler "graphic" method will suffice. However, note the discrepancy between the two methods at 3150 Hz! This example is based on gypsum board—a worst case. Still it shows that conservative designing is a good idea.

Table 8 *Surface Weights for Common Sheet Metal*

Gauge	Steel Thickness (in.)	Steel Weight (psf)	Galvanized Thickness (in.)	Galvanized Weight (psf)	Stainless Thickness (in.)	Stainless Weight[1] (psf)	Long Terne Thickness (in.)	Long Terne Weight (psf)	Aluminum Thickness (in.)	Aluminum Weight (psf)
10	0.135	5.63	0.138	5.78	0.141	5.85	0.135	5.64	0.102	1.43
12	0.105	4.38	0.108	4.53	0.109	4.57	0.105	4.39	0.081	1.13
14	0.075	3.13	0.079	3.28	0.078	3.25	0.075	3.14	0.064	0.90
16	0.060	2.50	0.064	2.66	0.063	2.60	0.060	2.52	0.051	0.71
18	0.048	2.00	0.052	2.16	0.050	2.08	0.048	2.02	0.040	0.57
20	0.036	1.50	0.040	1.66	0.038	1.56	0.036	1.53	0.032	0.45
22	0.030	1.25	0.034	1.41	0.031	1.30	0.030	1.27	0.025	0.36
24	0.024	1.00	0.028	1.16	0.025	1.04	0.024	1.02	0.020	0.28
26	0.018	0.75	0.022	0.91	0.019	0.78	0.018	0.77	0.016	0.22
28	0.015	0.63	0.019	0.78	0.016	0.65	0.015	0.64	0.013	0.18
30	0.012	0.50	0.016	0.66	0.013	0.52	0.012	0.52	0.010	0.14

[1]These are average weights for stainless steels. More exactly chrome-nickel stainless steels (series 200 and 300) have a surface weight of 42.0 lb/ft^2 1 in. thick. Straight chrome steel (series 400 and 500) have surface weights of 41.2.

Table 9 *Predicted TL in the Coincidence Range (dB)*

One-Third Octave Band Frequency	Graphic Method	Arithmetic Method	Difference	The One-Quarter points	
				Graphic	Arithmetic
1000	31	31	—	(1000)	(1000)
1250	33	32	1		
1600	34	33	1	1500	
2000	33	34	1	2200	2000
2500	29	33	4!		
3150	24	30	6!	3300	3000
4000	25	24	1		4000
5000	31	31	—	(5000)	(5000)

The difference between the two methods is most pronounced for gypsum board, because it has a sharper dip than most materials. There will be slightly smaller differences for other materials. However, even a 3 or 4 dB difference is enough to make you wonder which is the "right" method. Alas, neither is all that "right." The conservative idea—especially if the frequencies of the dip are critical—is to combine the worst features of both curves.

Damping imposed by the construction—fastenings, frames, gaskets or sealants—will usually reduce the plateau breadth for aluminum and steel. Typical constructions find steel with a coincidence region about nine bands wide.

SOMETIMES MORE ISN'T BETTER

Forgive a contrived example. Things like this *do* happen, however. Suppose a machine tool manufacturer has a fancy unit with an enclosed base which collects the chips or sawdust and the container for these (with dimensions of a few feet in each direction) is made of 12 gauge aluminum. He has had persistent complaints about the scream of the saw blade. It is a pure tone at about 2500 Hz, but he doesn't know that. He takes the attitude "Hang the expense! The new model is going to be quieter. Use ¼ in. aluminum plate for the chip bin!" (He doesn't know about coincidence either, does he?)

Let's see how this all works out: From the metal gauge table you can find that 12 gauge aluminum weights 1.13 psf and this leads (via the TL_{500} equation) to a TL of 22 dB at 500 Hz. If you plot the mass law line through this point with a slope of 5.5 dB per octave, you will project a TL of 35 dB at 2500 Hz. However, back there at a TL of 29 dB, coincidence began to lower the perfor-

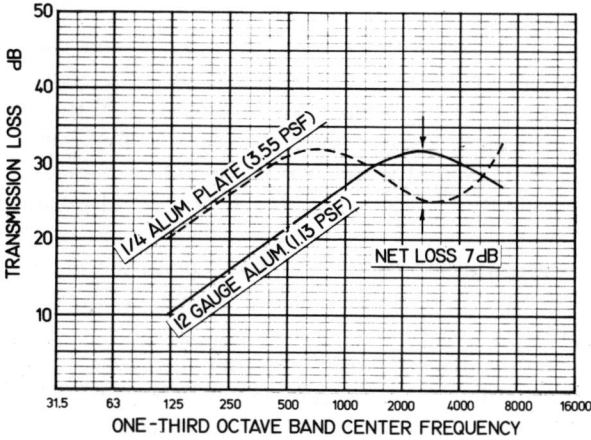

Figure 40 The machine tool manufacturer's mistake.

mance and if you plot that, too, you can see that the TL curve is peaking out at about 32 dB at 2500 (graphic method). Remember 32!

From the specific surface weight of 14.2 for aluminum, it is easy to find that ¼ in. plate will weigh 3.55 psf and from the TL_{500} equation, this translates to 32 dB of TL at 500 Hz. That's 10 dB better than 12 gauge was at 500! Now plot the curve and start coincidence at 29 dB as before. The TL for the expensive model is going to be about 25 dB—net loss: 7 dB.

Of course, most companies build a prototype before they tool up for production. The expense of even trying that design could have been avoided by the simple calculations we've just made. As a project, see what would have happened if he had switched to 16 gauge steel. (Answers in the neighborhood of 41 dB at 2500 Hz are correct and much cheaper!)

A BIGGER PICTURE

The two effects covered so far, *mass law* and *coincidence,* are usually the governing ones for single walls. But they are by no means the only ones. When you deal with structures "much smaller than a doghouse," for example, the mass counts for little and coincidence hardly ever has a chance to affect the performance.

Consider an extreme example: The plastic earmuff hearing protectors worn by baggage handlers and others around jet planes. If you suppose the plastic shell to be ⅛ in. thick and the plastic to behave like Plexiglas (an assumption, but good enough), you will be able to estimate that, based on mass law and coincidence, the earmuffs can't possibly work. You're right, of course. They work in a section of the TL curve where the *stiffness* controls.

Most of you have had a set of earmuffs in your hands. Could you break the plastic shell with you bare hands? Can you flex it? You cannot, because the material is inherently stiff, the shell is small, and the shape is a compound curve. Big question: If you can't flex it, do you suppose that the tiny pressure variation in the air we call a sound wave can flex it? Moreover, the earmuff is so small compared to the sound waves (for the usual frequency range of interest) that the shell is a point in space—there is no variation of pressure across its surface.

In the stiffness controlled region of the TL curve, the slope is about *minus* 6 dB per octave. For the frequencies you will usually be interested in, only small structures are stiffness controlled.

A large stiffness controlled structure (that doghouse again) is a dream as perennial as that of the perpetual motion machine. Let's not say it can't be done, only that when it is done, we will be able to design a cheaper and more effective wall even if it does weigh a little more.

The most common example of a large stiffness controlled noise control structure is the fuselage of a commercial aircraft.

Between the stiffness and mass law regions is a never-never land. Its name is the *panel resonance* region. Here, the wavelength of those bending waves, familiar from coincidence, is just right so that one, two, or three waves fit into the dimensions of the panel (wall) and are reflected back from the edges. The best image to explain what happens is a billiard ball on a frictionless table stroked off to rebound endlessly in a closed loop. The loop may be between the

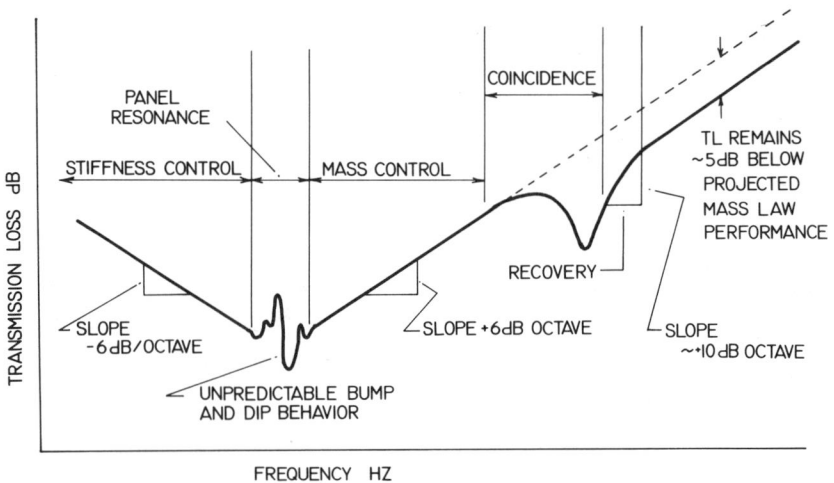

Figure 41 Small structures, or very stiff ones, can behave differently from what you would expect from mass law and coincidence alone. In usual constructions of "doghouse" size the stiffness and panel resonance regions lie below 50 to 100 Hz.

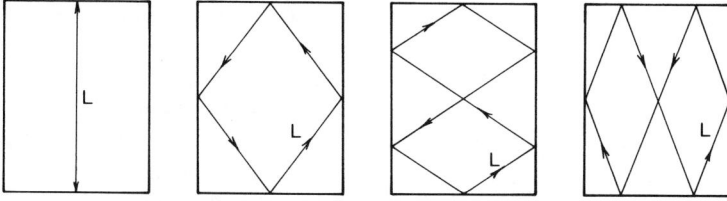

Figure 42 Whenever the length of one of the paths shown for a bending wave in the panel is exactly one wavelength long, a severe dip can occur in the TL curve.

ends or sides—or ends and sides in a diamond pattern—or figure eight pattern—or, but you get the idea.

The effect of this behavior is to make the TL of the wall change wildly in a small frequency range. You could probably contrive physical hardware showing a 20 dB difference in TL between adjacent one-third octave bands. The effect seldom is seen for usual materials in usual sizes for the usual frequencies of interest. Two cautions worth noting with regard to panel resonance are:

1 If you are working with small (a few feet) panels of thick steel or aluminum, either join them continuously to the structure at the edges so the bending wave won't be reflected or damp them heavily. This can be done after you discover the TL problem.

2 Do not expect mass law and coincidence to predict any useful TL data for a structure of two sheets of 24 gauge steel joined by rigid (foamed-in-place) urethane foam insulation. These panels are attractive in terms of cost, thermal insulation, and other characteristics. Acoustically, they are in panel resonance out to about 800 Hz. Acoustically, they are an abomination.

Once through panel resonance, the curve enters the familiar mass law and coincidence regions. After the coincidence dip has done its worst, there is a region that is usually called the *recovery region*. Recovery from coincidence is rapid. The slope of the curve here is about 10 dB per octave. However, the rise is limited, and when the recovery curve gets within about 5 dB of the straight line you can project from the mass law equation it comes no closer.

USEFUL TIPS IN ESTIMATING SINGLE WALLS

Normally mass law and coincidence are the two factors that will concern you most in predicting TL for a single wall. Real walls will have framing and perhaps other adjuncts, such as an absorber fastened to them. You need to account for the whole weight of the wall in finding mass law performance.

However, you base the coincidence performance on the sheet material of the wall.

The methods given here will work for typical light construction and also for heavy masonry. It is possible to think of constructions where the method would be in some trouble. For light sheets attached to massive framing, it is probably better to ignore the weight of the frame. If you were enclosing the area under a hopper by fastening sheet metal or plywood to the existing 4 in. channels, you would probably find a TL closer to that of the sheet alone than to that estimated for the whole structure.

Another problem construction would be such close spaced and rigidly fastened framing that typical coincidence behavior could not develop. The method does work for plywood or gypsum supported on normally spaced stud framing. It surely will not for plywood glued to wood honeycomb 2 in. thick with cell dimensions of 2 in. This leaves you with a gray area to worry about. At least, the area does not contain any common constructions.

EXAMPLE   What TL could you expect from a single sheet of $1/4$ in. plywood fastened over a 4×8 ft opening and framed only at the edges? How would this change if you stapled 3 in. of 1 pcf glass fiber blanket over it?

1   Find TL_{500} using

$$TL_{500} = 20 \log W + 21$$

Table 7 tells you that 1 in. thick plywood weighs 3 psf and $1/4$ in. plywood will, therefore, weigh 0.75 psf.

$$TL_{500} = 20 \log (0.75) + 21$$
$$= 18.5 \text{ dB}$$

2   Plot 18.5 at 500 Hz and draw in a mass law line with a 5.5 dB/octave slope.
3   Find the plateau height and breadth from the table. They are 23 dB and eight bands, respectively. Locate the beginning of the coincidence dip by finding where the mass law line reaches 23 dB (between 800 and 1000 Hz). Count off the eight bands ending between 5000 and 6300 Hz. Locate the quarter points of 1250–1600, 2000–2500, and 3150–4000.
4   Sketch in a smooth curve rising to 26 dB at the first quarter point and dropping back to 23 at the second. Continue it through 18 or 19 at the third and return to 23 at the end of the dip.
5   In tabulating your results, always read the nearest whole decibel below the curve. For the second part, find that 3 in. thick 1 pcf glass fiber will add 0.25 psf to the wall weight. Rework TL_{500} and find that it has risen 1.5 dB. Plot a new curve 1.5 dB above the first and tabulate results as before.

Your tabulated results would be close to the following data—allowing for differences in drawing and reading the plot.

Frequency One-Third Octave	TL of Bare Ply (dB)	TL of Ply with Fuzz (dB)
125	7 (11)	9
160	9 (12)	11
200	11 (14)	12
250	13 (14)	14
315	15 (16)	16
400	16 (17)	18
500	18 (19)	20
630	20 (20)	21
800	22 (22)	23
1000	24 (24)	25
1250	25 (26) **25**	27
1600	25 (27) **25**	27
2000	24 (27) **26**	25
2500	21 (26) **25**	22
3150	18 (25) **23**	19
4000	18 (23) **19**	19

The figures in parentheses are from an actual test of a 4 × 8 ft sheet of ¼ in. plywood. The figures in bold are predicted with the alternate "arithmetic" quarter points. The differences between the real and predicted TL are probably due to the sample size. The 4 ft dimension is small compared to a wavelength for lowest frequencies. The sheet is also too small to let coincidence do its worst. Except in the coincidence region, the difference between the actual and predicted TL is small and in the conservative direction.

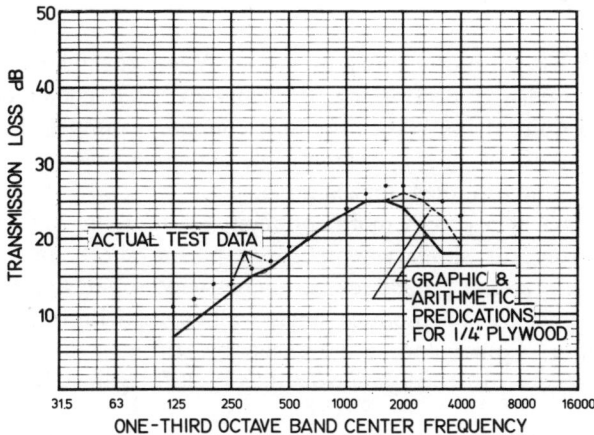

Figure 43 Graphic and arithmetic estimation methods compared to test data for ¼ in. plywood.

DAMPING (AGAIN)

When you intentionally—or even unintentionally—damp a single wall, you improve its performance. One trick is to use builders' foam tape, foam weatherstripping, or caulking between the frame and the sheet. This often produces a small improvement at higher frequencies, especially if the coincidence dip occurs there.

Damping goos and gunks can be used with varying improvement on light walls. A popular treatment producing excellent results is to laminate lead to a light sheet with contact cement. The composite or laminated sheets produced in this way cannot be reliably estimated by the method given for single walls. There is a way to estimate TL for them, but it is quite long and tedious. There are also questions about its reliability for some materials. The vast improvement in performance and the difficulty of predicting performance both stem from the same cause. The bending wave velocity in two materials like plywood and lead is vastly different. Thus the laminate is not easily brought into coincidence. The bending waves that do develop spend much of their energy in deforming the rubbery glue layer between the sheets.

For the combination of lead and plywood where the surface weight of each material is about the same (1 psf lead on 3/8 in. ply, for example), you can make a rough estimate that the plateau height will be about midway between those of either material alone. The coincidence dip will flatten so that the curve is always level or even rising with increasing frequency.

Laminated metal sheets that beat the coincidence effect by using a damping adhesive also are available commerically. Even where both layers are the same metal (steel or aluminum), they do better than a single sheet of the same metal for the same weight.

ODDBALL CURVES

The coincidence effect can produce two surprises. Where really large sheets of a single material are used with no framing, you can expect the coincidence dip to be quite sharp and quite deep. About the only common example of this sort of construction is in the use of glass in dramatically big windows—at some airports or modern office building lobbies, for example.

The other surprise arises from more commonplace building materials. Corrugated metal roofing, or similarly corrugated plastic sheet, or even metal building siding with recessed panels several inches wide separating the projecting panels of the same size, all produce single wall curves with *two* coincidence dips. Think about it. Coincidence is governed by stiffness. These sheets are relatively flexible across the corrugations but very stiff against bending along the corrugations!

In one patent noise control product, a plastic sheet has a deep sinusoidal corrugation with peaks about a foot apart. The two dips in that curve are about two octaves apart.

PROBLEMS

1   Worked problem. An 8×8 ft doorway inside the plant is no longer used and is to be blocked off by fastening ⅝ in. gypsum board to one side of a wood framework. The wood will by 2×4 studs 16 in. on centers. What TL can you expect if the wall is well sealed?

Data

Use Table 7.

Method

Find the TL using mass law and coincidence estimation methods.

Calculation

(a)   Find the weight of the gypsum sheet. From Table 7 the specific surface weight of gypsum is 4 psf at 1 in. thick. The sheet will weigh

$$(4 \text{ psf 1 in. thick}) \times \text{⅝ in.} = 2.5 \text{ psf}$$

(b)   Find TL_{500} and plot the mass law line at a 5.5 dB/octave slope:

$$\begin{aligned} TL_{500} &= 20 \log W + 21 \\ &= 20 \log(2.5) + 21 \\ &= 29 \text{ dB} \end{aligned}$$

(c)   Locate the coincidence dip. From Table 7, the plateau height is 31 dB and the plateau breadth for gypsum is seven bands. Your plot shows the mass law line reaching 31 dB between 630 and 800 Hz. The end of the coincidence dip will be reached seven bands later—between 3150 and 4000 Hz. Divide that region into quarters and plot:

Point	TL
Start	31 (the plateau height)
First quarter	34 (31 + 3)
Second quarter	31
Third quarter	24 (31 − 7)
Fourth quarter	31

Connect these with a smooth curve. (This is approximate, but close enough. Test data would show the curve crowding toward the right. You can divide the *numerical* differences between the frequencies for the beginning and end of coincidence by 4 and add that increment to the starting frequency four times if you want better to approximate position for the four preceding quarter points—the "arithmetic" method. The precision of either method is such that the simpler, graphic division is probably as reliable as the more complicated method.)

(d)   Now find how the curve is affected by adding the frame weight. Actual stud dimensions are 1.5 × 3.5 in. and in a running foot there will be

$$1.5 \times 3.5 \times 12 \text{ in.} = 63 \text{ in.}^3$$

Table 7 gives 2.4 psf for a 1 in. thickness, that is, 2.4 lbs for 144 in.^{3} Therefore, a running foot of stud will weigh

$$63 \text{ in.}^3 \times 2.4 \text{ lb}/144 \text{ in.}^3 = 1.05 \text{ lb.}$$

There are 5 studs, each 8 ft long or 40 running feet.

$$40 \text{ ft} \times 1.05 \text{ lb/ft} = 42 \text{ lb}$$

The wall area is 8 × 8 ft so the distributed frame weight is

$$42 \text{ lb}/64 \text{ ft}^2 = 0.66 \text{ psf}$$

To find the effect of the added frame weight, recalculate the TL_{500}

$$TL_{500} = 20 \log(2.5 + 0.66) + 21$$
$$= 31 \text{ dB}$$

(e)   The added weight of the frame increased the TL by 2 dB, so raise the whole curve by 2 dB. Where the curve indicates a fraction of a decibel, round the TL for that band to the lower whole decibel.

Answer

The results here might vary ± 1 dB in the straight line part of the curve and might be 1 dB above or below the extremes for the two methods of sketching the coincidence dip. *These are computational errors of the method.* Larger errors in comparison with real test data will exist, especially in the coincidence region. Gypsum walls often bottom out at 3150 Hz, for example. In practice be conservative in applying results of these projections.

Frequency (Hz)	TL (dB)	Frequency (Hz)	TL (dB)	Frequency (Hz)	TL (dB)
125	20	400	29	1600	32–35
160	22	500	31	2000	28–33
200	23	630	32	2500	26
250	25	800	34	3150	31
315	27	1000	34–35	4000	37
		1250	35		

2   Worked problem. What is the TL of a two-course brick wall 8 in. thick?

Data

Table 7.

Method

The difficulty here is that we are usually interested in TL over the range from 125 to 4000 Hz. This wall is so heavy that it is well into coincidence before 125 Hz is reached. There are too few data to check that the estimation method really works far outside the 125–4000 Hz range. However, pretend that it does and extend your plot to the nether regions of frequency so that you can predict the TL in the useful frequency range.

Calculation

(a) From Table 7 brick weighs 11 psf per inch of thickness or 88 psf in this case.

$$TL_{500} = 20 \log(88) + 21$$
$$= 60 \text{ dB}$$

(b) Project a mass law line with a slope of 5.5 dB/octave. It will intersect brick's plateau height of 37 at about 31.5 Hz.

(c) Develop the coincidence region curve for a plateau breadth of eight bands.

(d) Plot results and tabulate.

Answer

Frequency (Hz)	TL (dB)	Frequency (Hz)	TL (dB)	Frequency (Hz)	TL (dB)*
125	33	400	49	1600	63
160	34	500	52	2000	65
200	38	630	55	2500	67
250	42	800	58	3150	69
315	46	1000	60	4000	71
		1250	62		

3 Predict TL for a 6 in. low density cinderblock wall.
4 Predict TL for a ¹⁄₁₆ in. leaded vinyl curtain (1.5 psf).
5 Predict TL for ⅛ in. Masonite on 2 × 4 wood studs 24 in. on center.
6 Predict TL for ½ in. gypsum board on 2 × 4 wood studs 24 in. on center.
7 A few feet outside an alcove like that shown in problem 1 of Chapter 6 needs to be quieted to 84 dBA or less. The plan is to put a ¼ in. plywood wall across the opening to the alcove.

*Before flanking, of course.

Data

Assume that:

(a) All noise comes from the alcove.

(b) The plywood wall is to be framed with 2 × 4 wood studs 24 in. o.c.

(c) There will be no openings through this wall and the door and its frame will be at least as good as the plywood in TL.

(d) The reverberant build-up caused in the alcove by walling it in will be negligible. (Will it be?)

Octave band readings show:

Frequency (Hz)	L_p (dB)
125	81
250	87
500	94
1000	99
2000	97
4000	94

Part 1 What is the A-weighted level now?

Part 2 What will it be with the wall in place?

Answers

	Transmission Loss (dB)			
Frequency (Hz)	Problem 3	Problem 4	Problem 5	Problem 6
125	35	14	11	18
160	33	15	13	20
200	30	17	15	21
250	28	17	16	23
315	28	21	18	25
400	32	23	20	27
500	35	24	22	29
630	39	26	24	30
800	42	28	26	32
1000	46	30	27	34
1250	49	32	29	35
1600	53	33	31	35
2000	56	35	33	32
2500	58	37	35	26
3150	60	39	36	26
4000	62	41	38	31

7 Part 1 103 dBA.

Part 2 About 76 dBA.

9

ENCLOSURES, DOUBLE WALLS, BARRIERS

Much useful work can be done with simple single walls. However, they are not the only noise control tools based on transmission loss. Some of the others are explored in this chapter. The first of these is pipe lagging or close enclosures.

NEGATIVE TL?

For airborne sound, the TL of a wall may be very small but it is never less than zero. The transition region of panel resonance may never even occur. You won't find test data, but you can bet that testing a sheet of pasteboard would produce a curve that stays above zero. Panel resonance would not even be a problem because a material like paper is inherently well damped.

The bending waves in such a sheet would make the fibers rub over each other much like one of the common explanations of how a fibrous absorber extracts energy from the sound wave passing through it.

However, when the sound is not airborne, there can be a region where the "TL" becomes negative. "Not airborne" means that the "wall" is a wrapping or lagging applied to a surface radiating noise. Typically, this would be a noisy pipe wrapped in a glass fiber blanket and covered with an impervious sheet material like leaded vinyl. Here, some of the energy is surely applied directly and mechanically to the glass fiber blanket. It is transferred to the impervious cover. The "TL" in such an instance is called *insertion loss* (IL) and can be used in the same way TL is to predict what the noise level will be in the surrounding space. Where there is a major structural path for energy to get into the "wall" (impervious sheet), the rules all change.

There is often a region where the insertion loss is negative and, at some frequencies below 400 Hz (typically), the lagged pipe is actually noisier than the bare pipe. Another interesting characteristic of such a covering is that the insertion loss usually rises at about 9 dB per octave. Recalling that 6 dB per octave means taking the log of a square function, 9 dB ought to mean that something is happening as a function of the cube of frequency. Nobody has published

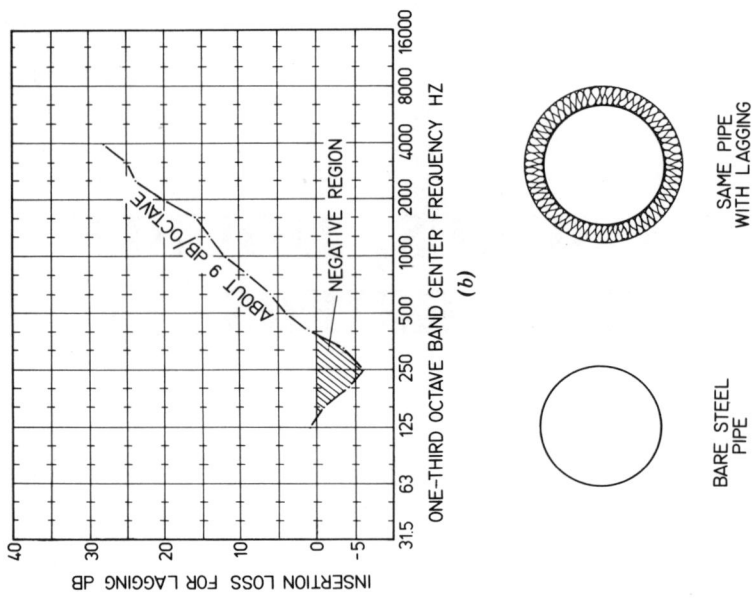

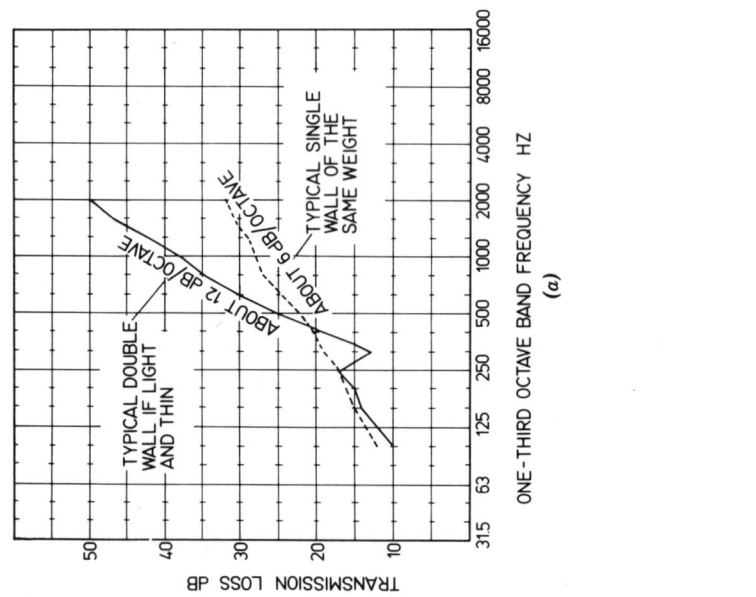

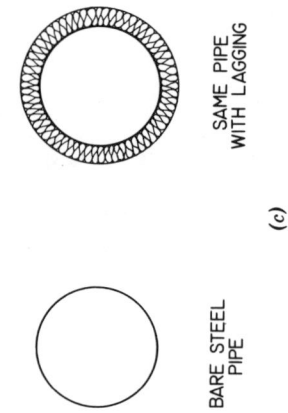

Figure 44 On the left, transmission loss curves for a single and a double wall of the same total weight. Close enclosure (lagging) produces the curve shown on the right for a related quantity called insertion loss. Under some circumstances insertion loss can be negative—the noise becomes worse for some frequencies.

(a) ONE-THIRD OCTAVE BAND FREQUENCY HZ — TRANSMISSION LOSS dB

TYPICAL DOUBLE WALL IF LIGHT AND THIN
ABOUT 12 dB/OCTAVE
ABOUT 6 dB/OCTAVE
TYPICAL SINGLE WALL OF THE SAME WEIGHT

(b) ONE-THIRD OCTAVE BAND CENTER FREQUENCY HZ — INSERTION LOSS FOR LAGGING dB

ABOUT 9 dB/OCTAVE
NEGATIVE REGION

(c) BARE STEEL PIPE — SAME PIPE WITH LAGGING

anything intelligible on that facet. However, acoustics is a young science and serious interest in lagging is only a few years old.

The data shown in Figure 44 are typical of 8 in. pipe with 2 in. of fuzz and a 1 psf covering. The negative region becomes smaller as diameter increases and disappears for large (>3 ft) pipe.

FAMILIAR GROUND

Thus for airborne sound there is no region where the TL goes negative. We have looked at the stiffness controlled region (either very small structures or very low frequencies) and the panel resonance region (not often troublesome for usual sizes and frequencies). What is next?

For that "doghouse" sized structure, or anything bigger, 50 to 100 Hz usually marks the point where mass law and coincidence are the controlling factors for airborne sound. We know about these.

When you think about it, you will realize that sometimes you will want a very high TL that mass law and coincidence can supply only at the cost of great, heavy, expensive walls. Is there a better way? Yes, as long as your TL requirements are not at low frequencies.

DOUBLE WALLS

The walls in your home or apartment and those in your office are most likely double walls. The familiar stud and plaster, or gypsum board wall, is a double wall and, although it is not built that way for acoustical performance, usually it does quite nicely. The typical rating (STC) for a typically installed double wall of gypsum board and studs is 25 to 35. That's more than enough TL for a situation where there are leaks around doors and often doorways with no doors in them. Higher TL in the wall would be no help at all.

Let's start to analyze the double wall by thinking of it as two single walls. If we could get single wall performance out of each face and if we were in the mass law region for each, we could expect the sum of their single wall performances—that is 6 dB per octave for each or 12 dB per octave. From the weights of each we could calculate the height of the curve. Simple? Too simple.

The 12 dB slope is approximately right. At least the analysis is headed in the right direction there, but the range of slopes you can expect from double walls runs from 9 to as high as 15 dB per octave. The height of the curve—that is, the TL at any frequency—*is not* the sum of the TL of the two single walls. Instead, it is the TL you would expect from a single wall of equal weight.

It may help to look at it this way: When we said that a double wall was two single walls, you might have asked yourself "Single walls of what?" Single walls of a room? Ah, then you *can* expect to get something like the sum of the single wall performances. In other words, if the only way sound can get from room A

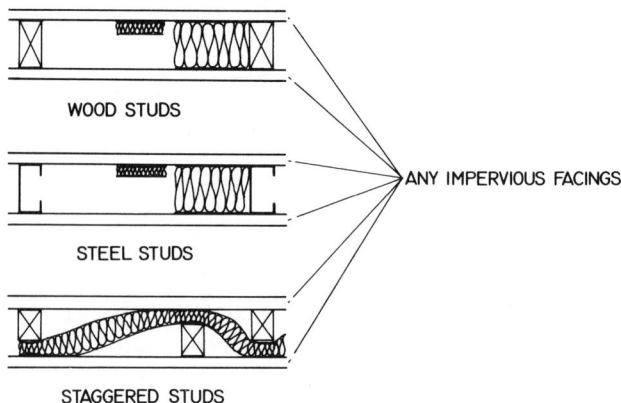

WOOD STUDS

STEEL STUDS

ANY IMPERVIOUS FACINGS

STAGGERED STUDS

Figure 45 A few of the variations on the double wall theme. The only two basic requirements are faces that are impervious (e.g., *not* perforated metal) and enough cavity space to allow the faces to act independently. The amount of fuzz may vary from none to a lot.

to room C is by going through a single wall into room B and another single wall between rooms B and C, then, indeed, the TL will be the sum of the single wall TL.

There is a minor problem in room B. If it is quite hard (reverberant) the actual noise reduction between rooms A and C won't be quite as good as the sum of the TL of the two single walls.

A double wall doesn't do this well for two reasons. Typical spacing between the two faces (sometimes they are called wythes) is a few inches and usually the air is trapped in narrow cells between studs and other frame members in this space. Although air is compressible, for the scant volume at play here and the very brief time allowed between successive pressure fronts of the sound wave, the air is stiff enough to couple energy to the second face fairly well. In some double walls the framing is quite light and stiff and the wall faces are fairly heavy by comparison. In a perfectly ordinary gypsum board wall, for example, the faces weigh 2.5 psf and the stiff steel studs weigh approximately 0.5 lbs per running foot. Here, the structural coupling between the faces makes the total wall act like a single wall part of the time and, in the high performance part of the curve, it permits a slope of only 10 dB per octave.

YET ANOTHER RESONANCE

In the TL curve of a double wall, there is a frequency absolutely critical to the performance that is usually called the *cavity resonance* frequency. (Actually, this name is inappropriate, because it has already been taken by another effect with a much better claim on the name. If you read an article that talks about *mass-air-mass* resonance, know that its author has been properly careful in his language. It is *mass-air-mass* resonance we are now looking into—though everybody calls it "cavity resonance.")

Think of the air trapped between the two faces of a double wall as a spring and the two faces themselves as weights attached to the ends of the spring. This system is peculiarly resonant. The two faces tend to fly in and out quite freely

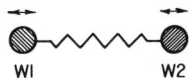

Figure 46 A spring with a weight on either end is a good analogy to the mass-air-mass system.

for the particular frequency that suits a set of requirements. We have already run into that set of requirements for the special case where both of the faces have the same weight, W in psf. Remember the panel absorber? (The form is right!)

$$f = 240/\sqrt{Wd}$$

where f = resonance frequency (Hz)
W = weight of the faces (psf)
d = spacing between faces (in.)

For one reason or another, though, the two faces of the wall may be quite different. In this case: (Even the constant is right!)

$$f = 170/\sqrt{\frac{W_1 W_2}{W_1 + W_2} \cdot d}$$

where f = the resonance frequency (Hz)
W_1 = the weight of one face (psf)
W_2 = the weight of the other face (psf)
d = the spacing between them (in.)

This is a simple enough equation to solve. Sometimes, though, if you get into serious design work, you will be faced with a problem. If, for some reason, one of the faces has to be a certain thickness of a certain material and you need to find the weight of the other face to set a certain mass-air-mass resonance, you will find the trial and error calculation quite tedious—or you may resort to the handy nomograph (Figure 47).

It lets you work back from your preferred frequency to the weight of the other face quite easily. Sometimes it will even save more work than that. It will tell you "there ain't no such beast!" Try the nomograph out. Suppose you were asked to design a wall with a 2 in. cavity where one of the faces had to be 1 psf and the cavity resonance was to be 100 Hz. Use a triangle or sheet of paper to line up the 2 in. cavity (right-hand scale) with your requirement of 100 Hz and put your pencil point on the blank line between the two exponential curves. Pivot the straightedge until it runs through the face weight of 1 psf and watch it completely miss the other face weight scale!

If you pivot the straightedge back, it will tell you that the lighter face has a minimum weight of 1.5 psf. With more juggling, you could establish that if the 1 psf figure is the controlling requirement, you will need a minimum cavity of $2^7/8$ in. (of course you would use at least 3 in.).

ESTIMATING A DOUBLE WALL

The wall being estimated here is composed of sheets of $1/2$ in. gypsum board fastened to either side of 2×3 in. wood studs (1.5×2.5 in. actual dimensions) with 2 in. of 1 pcf fuzz in the cavity.

The start of this exercise, at least, is familiar: Find the total weight of the wall in psf! Apply the mass law equation to find the mass law TL at 500 Hz ($TL_{500} = 20 \log W + 21$) and plot the point. Run your favorite slope through it, but draw the line lightly because this time many adjustments will be required.

Calculate the mass-air-mass resonance frequency and draw it as a light vertical line through the mass law line. The resonance will extract a 6 dB penalty—the actual TL for this frequency will be that much worse than mass law. Actually, the penalty may be as little as 2 to 4 dB, but it takes a nice judgement to say when. The next problem is where in the world is 107 Hz (or whatever you calculated) on this paper? Don't worry. Make allowance for the ever compressing scale of log paper, but also make allowance for the fact that the weights weren't very exact either, nor, for that matter, was the equation. You will put the pencil down in about the right place.

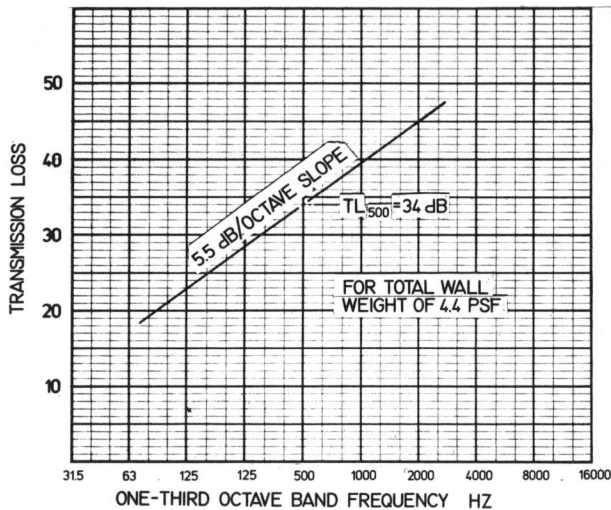

Figure 48   The first step in estimating TL for a double wall is to establish a mass law line based on the total wall weight. This will include the weight of both faces, the framing, and any fuzz used.

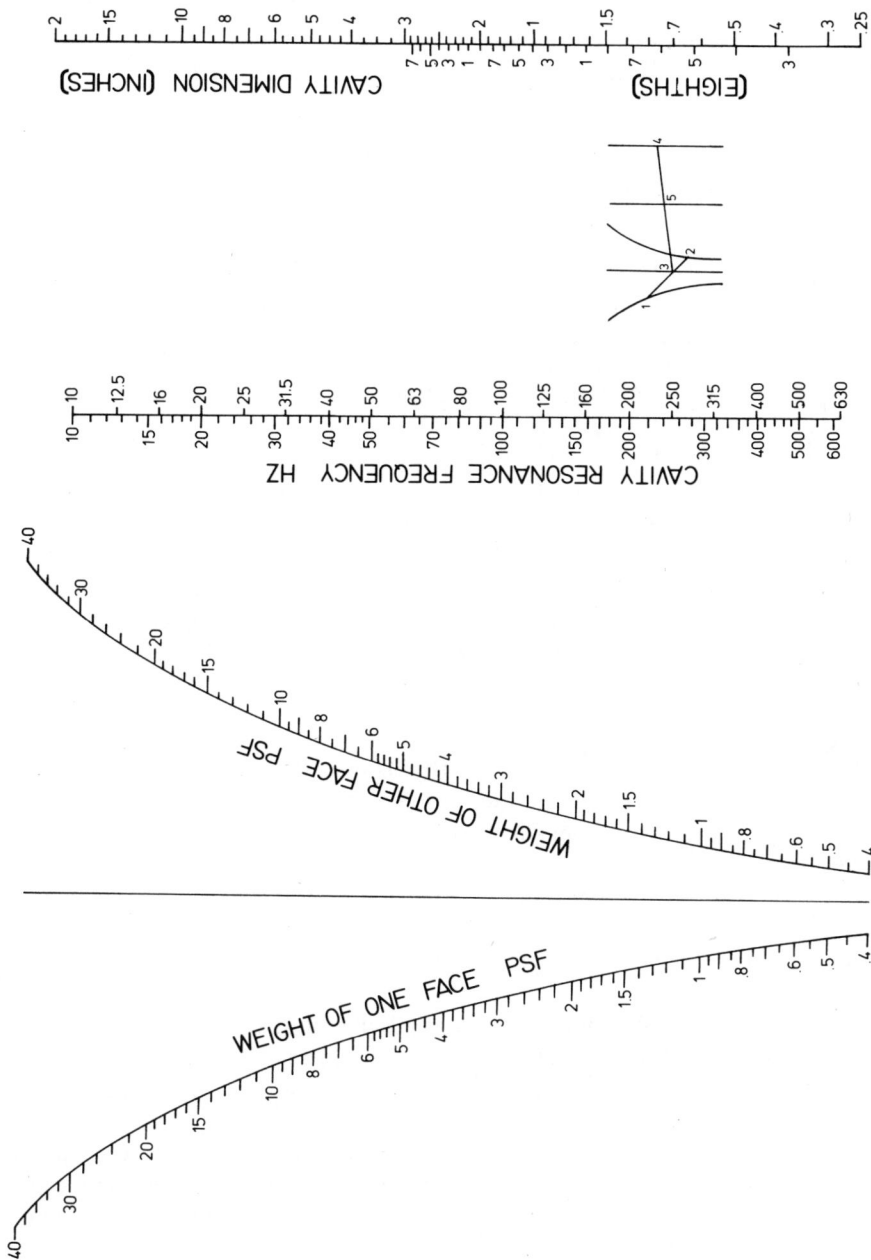

Figure 47 The mass-air-mass resonance nomograph is faster than the calculation when the two faces are of different weights. Find the weights of the two faces on the two curved scales and align them. Then pivot the straightedge on the point located on the straight line between the weight scales until it lines up with the cavity dimension. Read the mass-air-mass frequency directly.

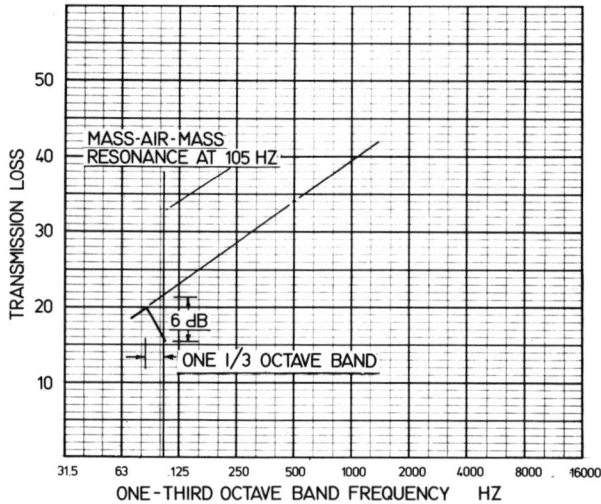

Figure 49 When you have found the mass-air-mass resonance frequency locate it on the plot and sketch in the beginning of the cavity dip. The 105 Hz resonance frequency shown was found by nomograph. A more precise figure of 107.5 could be calculated. Finding the difference between those frequencies in the plot would be difficult.

Mark the spot on the mass law line that is one-third octave band below the resonance frequency and darken the mass law line up to this spot. Then turn right and darken the line from there to the bottom of the cavity dip.

Now return to the intersection of the cavity resonance frequency line and the mass law line and project a new line that rises at 12* dB per octave from there to the top of the paper (lightly, of course).

Now go back to the two original faces of the wall and plot their mass law lines.

Correct them for coincidence. If you are doing all this on one sheet of paper, neatness counts! Pick off the difference between mass law and the single wall TL curve after coincidence for each of the faces and subtract their total, graphically, from the light line you have just projected from the intersection of the mass law line for the double wall and the cavity resonance point. (There is no law that says coincidence cannot begin at a frequency lower than cavity resonance. In such a case, subtract the coincidence correction from the whole curve you have developed so far.)

There is a little bridge we have left out. If you are lucky and all the coincidence dip is well beyond the cavity dip, you can now find a point on the 12 dB line that lies 2.3 one-third octave bands above the cavity resonance and connect the bottom of the cavity dip with that point.

*Twelve decibels per octave slope is the theoretical, and provisional, slope. Use it for now. Some following notes will help you decide what slope this last line really ought to have.

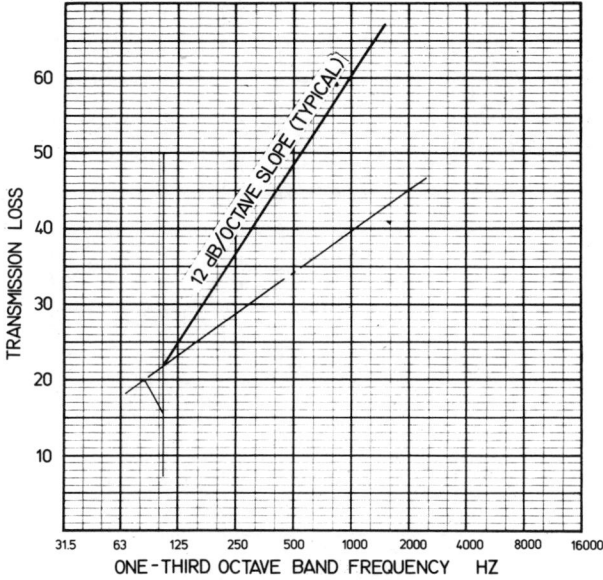

Figure 50 The next step is to project a 12 dB ($\pm$ see later section) per octave line from the intersection of the mass law line and the mass-air-mass resonance frequency.

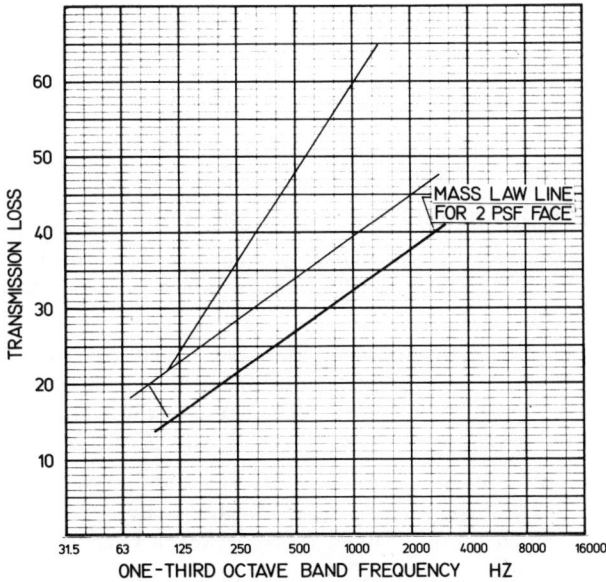

Figure 51 You must now go back and plot in the mass law line for each of the faces. In this example they are identical.

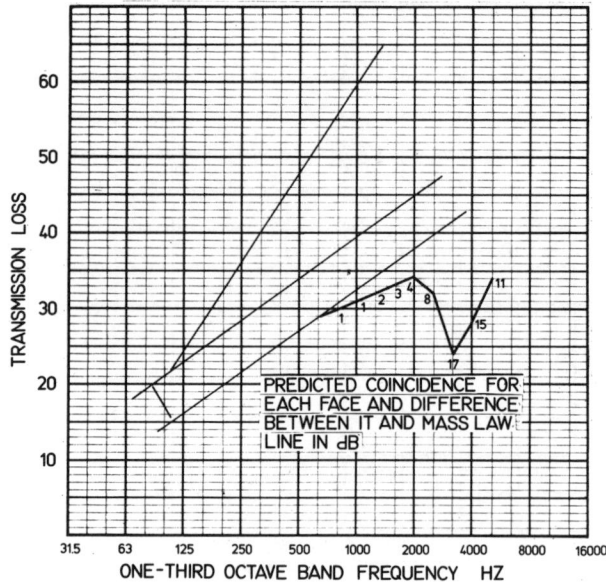

Figure 52 Find the curve for the coincidence region for each face. Here both faces are the same thickness of gypsum board, and the arithmetic method has been used to find the coincidence dip.

Voilà! You have the curve—or is this a bad joke? You can be assured you do and that we have taken a lot of shortcuts to get the answer so easily. You can also be sure that the result is wrong if it predicts TLs of 55 to 65 dB at some frequencies, because we have not even considered *flanking*.

FLANKING

Transmission loss is the figure of merit for walls. Walls are defined as surfaces "sealed on all edges so that there is no important way that sound on one side of the wall can get to the other side." The "unimportant" way implied in that definition is flanking.

Flanking is any other path sound takes to move from one side of a wall to the other. The most serious flanking paths are through air leaks in the wall. Openings from doorways to cracks are important. Where the wall is really hermetically sealed, there are still other flanking paths—those studs between the gypsum board faces, for example. This is usefully called structural flanking and means that some parts of the wall (those connected to heavy structural members that are transmitting structural vibration, for example) provide preferred paths for sound to travel from one side of the wall to the other.

The lowest blow, if you are looking for a TL of 100 dB, is that your structure

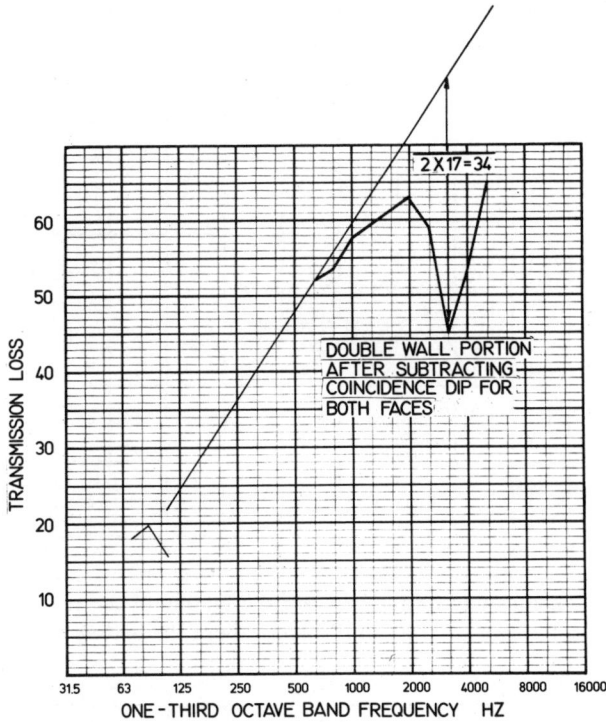

Figure 53 Graphically subtract the difference between the face mass law and coincidence TLs from the 12 dB/octave line. Here it has been done twice to account for both faces.

will be footed on earth. Believe it or not, if you utterly isolated one face of your double wall from the other, but let them both be supported by mother earth, the energy coupled into the first face from the air would go down into its footing and foundation, into bedrock, back up by the same path, and finally show up on the quiet side of the wall. The noise reduction by that path does not exceed 80 dB! You don't have to rule out noise reductions of 100 dB. But you do have to start looking at many flanking paths.

HOW SERIOUS ARE LEAKS?

The predominant flanking path is through air leaks in a wall. These are *important* when the noise reduction you seek is small. The effect of flanking through air leaks when you need high TL is *overwhelming*.

As you might expect, the effect of leaks can be calculated. In fact, the same method lets you estimate what will happen in any real-world wall that has win-

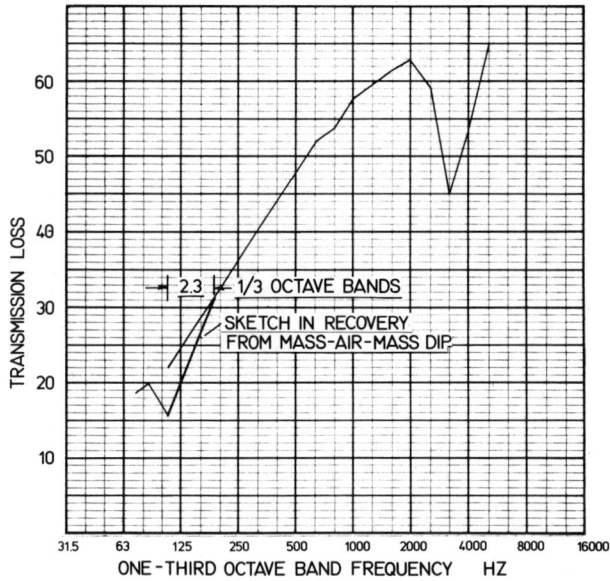

Figure 54 Draw a connecting bridge between the bottom of the cavity dip and the 12 dB/octave line at about two and one-third bands above the bottom of the dip.

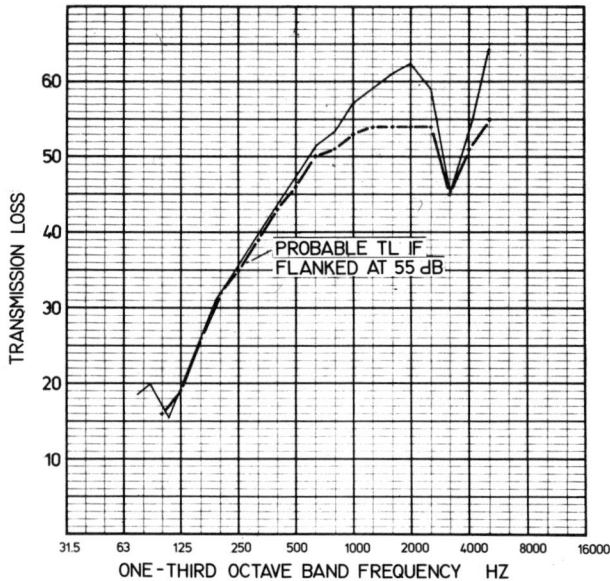

Figure 55 The ideal TL performance must be corrected for flanking. Here, based on experience, it is assumed that TL will never exceed 55 dB. That would be a typical figure for a well-built, well-sealed, unpenetrated wall of this type. Where the predicted performance is not an integral number, round downward. For the 35.5 dB at 250 Hz, for example, report 35 dB.

142

dows, doors, or different materials of construction for different parts of the wall. The equation runs

$$\text{TL}_{\text{eff}} = -10 \log\left[\frac{S_1 \cdot 10^{-(\text{TL}_1/10)} + S_2 \cdot 10^{-(\text{TL}_2/10)} + \cdots}{S_1 + S_2 + \cdots}\right]$$

where TL_{eff} = effective overall transmission loss of the wall for the frequency considered (dB)

        TL_1 = transmission loss of the first component of the wall for the frequency considered (dB)

        S_1 = surface area of the first component of the wall (ft^2) [and remember that leaks (cracks and gaps) are components of the wall, too]

You may have noticed that the numerator ends in dots—any number of components can play. Also you may have noticed that this equation needs to be solved for each frequency of interest (typically, for each one-third octave band).

You must admit that running this whole mess out is going to be a tedious affair. If it were often required, we might let a computer do it. Because it is not used so often, you can resort to a nomograph (Figure 56).

LOOSE ENDS 1: STC

Like the noise reduction coefficient encountered earlier in absorption, STC is quite useful to architects and is sometimes misleading in industrial noise control. STC is a complicated compromise reached in the middle 1960s as a way of producing a single number that is supposed to tell you the merit of a wall in TL.

A little history may be helpful. TL data such as those published through the 1950s were found with one-half octave band resolution and the TL at 11 of these specified frequencies were numerically averaged to produce the (surprise!) "11 frequency average." Then, in the early 1960s, one-third octave band resolution for TL tests became fairly standard. Sixteen of these were tested.

Instead of adding more bands to the number to be included in the average, the bands sampled were reduced to nine (skipping some at higher frequencies) to produce, as you have guessed, the "9 frequency average." By giving more importance to the lower frequencies, the "9 frequency average" generally tended to give a greater numerical distinction than its predecessor between any two walls tested, since, as you have seen, TL at the lower frequencies is almost always lower than at the high end. Moreover, the high end in those days often *tested* about the same for any wall because of flanking in the test labs.

The first little moral to be drawn from this history lesson is that if you are given data with an 11 frequency average or a 9 frequency average, you can safely discard it out of hand. It is from another age of acoustics and not generally reliable.

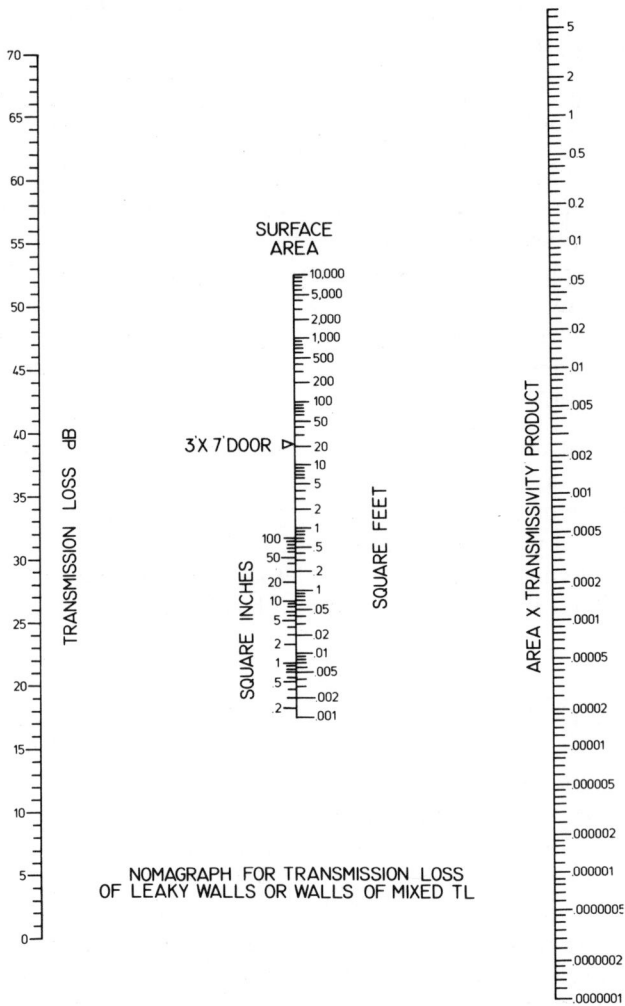

Figure 56 This nomograph can be used to estimate the effect of leaks (TL=0) or other elements of a wall (doors, windows, etc.) with TLs different from the main part. For each element align the TL and the area of the element to read the product on the right-hand scale. It is power. All the powers can be added (a separate operation on paper or with your calculator.) Reenter the nomograph with the total power on the right scale and align this with the total area on the center scale. The effective TL will read directly on the left-hand scale.

To go back to the history. The 9 frequency average proved unreliable as an indicator for customer satisfaction because it was used to specify walls between rooms, offices, and so on, where the big job was to make speech from the adjoining room unintelligible. If everything worked out just wrong, monster dips in the curve caused by coincidence could be fitted in between the few sampling points of the 9 frequency average at high frequencies. Walls with impressive numbers might leak intelligible speech like a sieve, while more modestly rated ones did a fair job.

By the mid-1960s, the acoustics committee of ASTM decided that something had to be done. They developed the sound transmission class (STC) idea and it was a good one. Bowing once again to Fletcher and Munson, they developed a curve with which to compare any TL test curve. The idea was that the test data might only fall 2 dB below the STC curve at any point and not at all below it in the critical frequency range. You might well wish that we still used that curve.

That was the "old" STC curve. Today we have a "new" STC curve. It is still recognizable as an adaptation of the Fletcher-Munson family, though a trifle different from the old one in its inflection points and slopes. The disappointing feature of the "new" STC is not the curve but that the TL test curve may fall an *average* of 2 dB below for all points except that no single point may be more than 8 dB below it. More than a decade has passed since the new STC was adopted and perhaps the rancor of the arguments then has all passed, too. As soon as the new STC was adopted, the General Services Administration (which does all the government's specifying and purchasing) raised the STC requirements on all gypsum board walls by several decibels. Acoustics is not all physics, math, and purity.

In industrial noise control, you won't have much use for STC anyway. Unfortunately, it is sometimes the only number reported for TL for a material or construction. In the roughest way you can take STC to be something like the TL at 500 Hz. You can be assured that if an STC exists, a TL test curve does also—so a letter requesting the curve may help. The problem with STC is the same as that with NRC as a figure of merit for absorption. Both are single number ratings developed for use in *architectural* acoustics where the problems, certainly the noise sources, are often quite different from yours.

LOOSE END 2: THE EFFECT OF ABSORPTION IN A DOUBLE WALL

If you can remember the beginning of the discussion about the differences between a single wall and a double wall, you will remember that there was a problem with the reverberant build-up in room B. Even in a discussion of a simple double wall, reverberant build-up between the faces is a problem.

One way to subdue reverberation, of course, is with an absorptive material—typically an inexpensive glass fiber batt, in this case in the space between the two faces. Considering that the source of noise in the space between these faces is the noise radiating from the inside of the first face and the admonition so frequently repeated in the absorption section, you might think that absorption is never effective near the source and think that an absorber between the faces could not help the TL. You would be wrong.

The space between the faces is typically 4 in. and often much less. To the extent that a sound wave can be developed in this cavity, it is bound to be reflected back and forth many, many times per second. What if there is an absorber there to be traversed at each reflection? Indeed, the inclusion of an absorber between the faces is a very good idea.

Theory predicts that a double wall ought to have a slope of 12 dB per octave. *In general,* if you put some sort of absorption in the cavity, it will. If you do not, the worst case is a double wall with a slope of 9 dB per octave.

Now gypsum board is an interesting material in this regard. It comes furnished with a paper face—and not such good quality paper at that. The face is somewhat open and somewhat fuzzy—though hardly what you, as an acoustician, would choose as an absorber. If you build a gypsum board double wall, you can probably count on a 10 dB slope in the double wall region. If both faces are hard ("Masonite," sheet metal), think 9 dB per octave.

When you have done some of these calculations, you will come to appreciate the vast performance difference that results from a seemingly trivial difference in the slope.

Now, magic! If you choose the absorber to have just the right characteristics, you will get a 15 dB per octave slope and also minimize the dip in the cavity frequency region. The bulk density of the glass fiber blanket that does this will be between 0.75 and 1.25 pcf and its thickness between about 1.1 and 1.25 times the cavity it fills. This means that, for a 1 in. cavity, you put up the frame and one face, and staple in $1\frac{1}{4}$ in. of 1.25 pcf glass fiber blanket, for example, and then fasten the second face in place on the frame, compressing the glass.

There are no promises here. It sometimes doesn't work out to be a 15 dB slope and it won't ever do better. But in this range, you are getting as much as possible from a double wall. Things that help are wide spacing between the framing members (studs) and cavities thicker than an inch. The notion is that the mechanical energy of the first face is partly coupled, mechanically, into the glass fiber blanket and there it is dissipated by friction of fiber-over-fiber.

OTHER TRICKS

Coincidence, too, can be brought under control when the faces are thin and stiff. The trick here is to add limp mass (e.g., lead, leaded vinyl, heavy mastics) that contributes nothing to the stiffness of the laminate. The light and stiff layer of the laminate wants to go into coincidence at a low frequency and the heavy and limp member wants to go into coincidence at a high frequency. Like two mules pulling in opposite directions, neither is satisfied and the dip associated with coincidence can be essentially dispensed with if the weight of the two layers is comparable. If you use lead sheet, laminate it to its substrate with a viscoelastic (contact) cement.

This surely is no area for an apprentice acoustician. The math is not especially trustworthy and it is complex and not wholly published. It does work and selected test results by the manufacturers of heavy and limp materials may be of some help to you.

Structural flanking can be allayed, if not defeated, in double walls, by means of resilient clips that support one or both faces of the wall without rigidly attaching them to the framing. These are commonplace in commercial and institutional construction and a competent and honest contractor* will know just how to undertake a job like this for you. You can tell whether he's competent, at least, by trying out the things you've learned here about TL of walls. Don't worry about whether he knows *all* of them—listen to him to see if he has the general idea. And don't expect much sense in the question of coincidence.

For a little more money, you can use staggered studs. These do wonders with structural flanking. Both the latter schemes, though, can only help at high frequencies and extreme reaches of TL. Don't use them willy-nilly: They cost too much for that.

BARRIERS

Until now leaks have been treated as though they were small areas like cracks in a wall or an open doorway in a big wall. A barrier was defined at the beginning of Chapter 8 as an obstacle to sound's travel but an obstacle where the noise was free to go around one or more edges. Barriers can be quite useful. They work best outdoors. There they can reduce the noise traveling *in some directions* as much as 24 dB.

Indoors, reflecting walls and ceilings will ensure that noise that would otherwise escape bounces into the space behind the barrier. It can be quite tricky to estimate how effective the barrier will be under these conditions.

One break that Mother Nature gives you with barriers is that their performance increases with frequency and also with their size. This means that you are usually dealing with a small noise reduction because the frequency is low or, if you're near the 24 dB limit, the barrier will be big and hence fairly heavy. In either case the TL of the barrier—that is, its ability to stop noise from going "through" it—will be so good compared to what gets around it that you will not have to worry about or make any correction for the TL.

The best way to think of how barriers work is to think of them as casting shadows: *sound* shadows, of course.

Now think back to our imaginary experiment with the firecracker in the box.

---

*On the question of honesty, Kipling said: "Who can doubt the secret hid/beneath the Cheops pyramid/is that the contractor did/Cheops out of several million."

The hole in the box casts a nice sharp shadow for the flash of light and not much of a shadow for the noise, which tends to move away from edges (or holes *or* barriers) in spherical shells.

If you were to compare the wavelengths of light and sound, you'd find light has a much smaller wavelength than sound of any frequency. Does short wavelength have something to do with barrier performance? Indeed.

A practical outcome of this is that the performance of any barrier in decibels of noise reduction always increases with frequency (because the wavelengths are shorter). The dimensions of the barrier and the paths sound takes from the source to the receiver need to be reckoned in wavelengths. The fact that the basic effect at play is diffraction introduces a hyperbolic function. In all, the mathematical basis becomes so complicated that there is little point to attempting to see why the relationship works out as it does. However, when you need to estimate the NR available to you from a barrier, use the Maekawa equation

$$NR = 20 \log\left[\frac{\sqrt{2\pi N}}{\tanh\sqrt{2\pi N}}\right] + 5 \text{ dB}$$

where $N = (F/565)(A + B - d)$
    NR = noise reduction (dB)
      F = frequency (Hz)
  $A + B$ = shortest path length around the barrier (ft)
      d = straight-line distance, source-to-receiver (ft)

LIMITATIONS FOR BARRIER CALCULATIONS

People can, and do, use the preceding equation to estimate the effect of a barrier on noise level. The solution of the equation is embodied in the nomograph in Figure 57. There are many other factors to be taken into account in the real-world performance. The more important ones follow.

Reflected noise will lower the performance of the barrier to less than your estimate. The very worst situation is to use a barrier indoors, because the noise not only diffracts around the edges but bounces off walls and ceilings. Even out-

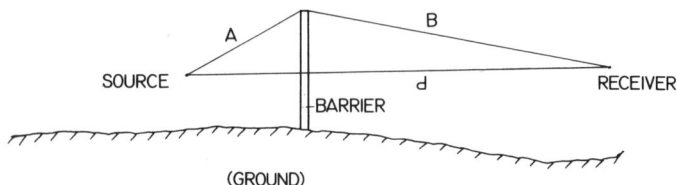

Figure 57   The key dimensions in reckoning a barrier's effectiveness are the straight line path (*d*) between the source (S) and receiver (R) and the shortest path around the barrier (*A* and *B*).

DIFFERENCE IN PATH LENGTH (A + B − d) IN FEET

.5 .3 .8 1 1.25 1.6 2 2.5 3.15 4 5 6.3 8 10 12.5 16 20 25 31.5 40 50 63 80

FREQUENCY HZ

8000 4000 2000 1000 500 250 125 63 31.5

NOISE REDUCTION (ONE EDGE) BY BARRIER dB

24 22 20 18 16 14 12 10 8 7 6

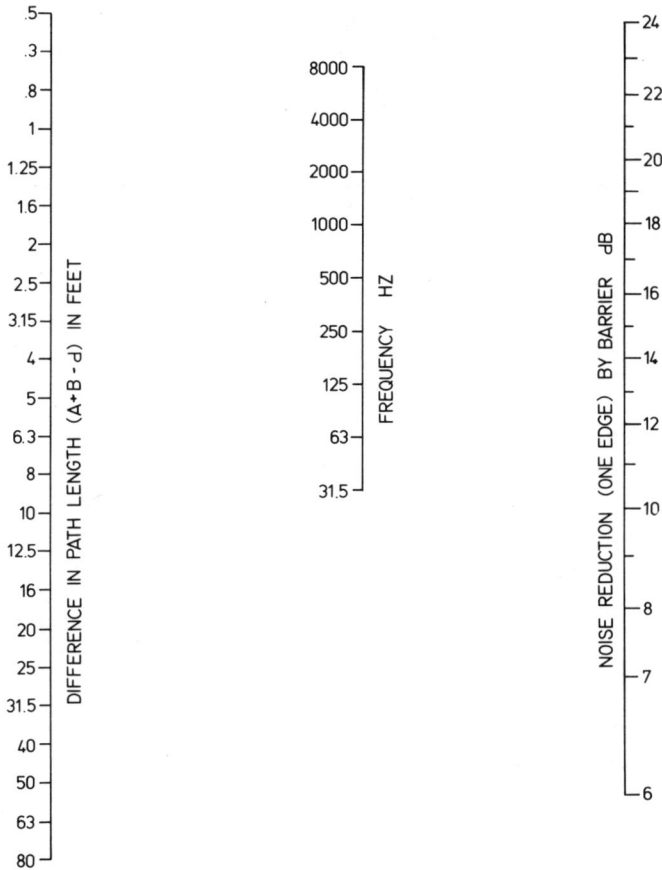

Figure 58 Nomograph to estimate the noise reduction afforded by a barrier. It considers one edge only and assumes that there are no other paths.

doors the presence of large buildings can lower barrier performance if they are in critical locations.

Unless you consider all the free edges of the barrier, there will be more diffraction paths than you have taken into account. The limit of performance for any one edge of the barrier is given variously as 20 to 24 dB. For a freestanding straight wall reaching a limit performance of 24 dB on all three free edges, the actual noise reduction should be about 19 dB, therefore, and not 24 dB.

The equation (and nomograph) considers point sources. This is a good approximation for the case of a fan discharge duct 18 × 24 in. when you are building a barrier 12 × 30 ft, for example. However, what if the noise source is a big transformer 10 × 10 × 12 ft, for example? Or a cooling tower three stories high? You will have to consider extremely large barriers to make these look like point sources. The usual approach to large sources is to integrate over the source surface, as though it were many, many point sources and taking hyperbolic tangents, square roots, and logs for all frequencies at each point. Fun? You simply don't have enough time to do it. Many acoustical consulting firms have written computer programs to do the job. If you are familiar with computers, you may be impressed that a good sized computer will churn for 30 to 45 sec after receiving the last datum and being told to find the answer.

Special effects can produce performance better than estimated. If your barrier is a building with a width comparable to or greater than its height, you'll get more noise reduction than the equation predicts. If the barrier is absorptive, or the ground is, performance is better than predicted.

HYPERBOLIC HELP

Hyperbolic functions are used less than once a week by most engineers. No doubt, there is a handbook gathering dust on your shelf that will give them, or you may be fortunate enought to have a calculator that produces them on order. But if you are working one of these problems on a plane or on Sunday at home with the wrong calculator and no handbook, your refuge in desperation is

$$\tanh x = \frac{e^{2x} - 1}{e^{2x} + 1}$$

where *e* is Napier's constant (2.7183 approximately). You'll also find hyperbolic tangents in Table 22 in the appendix of this book.

ENCLOSURES

Boxing in a noisy machine (or its operator) is sometimes attractive as a noise control measure. There are always two sets of problems you must face when you consider enclosing a noise source (or receiver). The acoustic set is often not nearly as great a problem as the practical set. We'll concentrate on the acoustic problems here, but:

Can you get the feed into and the product out of the enclosure without letting all the noise out, too?

Be sure you ventilate or cool heat producing equipment. What's good for containing noise is good for containing heat!

If you don't provide for *easy* access for maintenance, the maintenance man will ... with a fire ax!

Will you need some indication of malfunction of equipment like an alarm or at least a window (and lighting inside)?

Finally, before you broach your idea for an enclosure to the production manager, reread Dale Carnegie's *How to Win Friends and Influence People*.

If the enclosure *still* looks good, here are some helpful acoustical tips. Sketch an enclosure embodying all the practical requirements and apply your knowledge of TL and flanking through the air leakage paths to see whether you have met the noise reduction requirement with at least 10 dB to spare. If you have, you have probably solved the problem. If you don't meet the goal with 10 dB to spare, look into absorption inside the enclosure.

In fact, if you do many enclosures, looking into absorption on the first calculation will become habitual. Suppose, for the sake of visualizing what happens, we say the enclosure is to be a 1 ft cube. When there was no enclosure, five of the sides of the cube were empty space with an alpha of 1. If the enclosure is to be plywood (assume alpha = 0.1 for this example), the L_p inside the enclosure will have gone up about 10 dB compared to the L_p before enclosing. That's why we added 10 dB to the goal.

Now suppose, after playing with absorption, you find that any cheap, convenient materials of construction (including absorbers) fail to give you the reduction you want. You find that increasing the TL of the walls doesn't help because the air leaks for feed and product (and scrap) are controlling. You could consider adding mufflers to these openings. (Mufflers will be discussed in chapter 10.) You might also consider making the enclosure larger.

This can do wonders. If you make it about three times as large (in linear dimension), you not only bring more absorption to bear, but you have let those beautiful spherical waves expand and be diluted. (By almost 10 dB, if the source had been a point source. After all, $3^2 = 9$ and 9 is about 10, for very large values of 9.)

NOISE CONTROL SCHEMING

The happiest situation is a case where the source of noise is physically small and you can put a big enclosure around it. You will be able to add absorption to the inner surfaces and take care to tune it well to the source spectrum. (Put your big alphas where the noise is.)

The next best case is a simple source like a fan or pump that can be totally enclosed (except ducting and piping) because the stuff flowing through the operation is already contained. Access will be needed only for maintenance. Vision ports will probably not be required. If you can keep the motor outside, usually you will not even have a heat problem.

Next best is the same operation with the motor inside. It needs ventilation.

However, the most common situation is a punch press or a planer with lots of feedstock, product, and scrap and a maintenance man or adjuster flying in and out. While the machine making noise is big the actual source or sources of noise may not be. To the extent that you can put a big enclosure around a planer, for example, try to arrange things so that the first and last cutting operations are as far as possible from the inlet and exit openings.

Does that mean that you should design the enclosure to be long and tunnel-like? Long, yes. Tunnel-like, maybe. Acoustically, making the enclosure as big as a barn would be helpful—but you can't: that's too expensive. Your interest lies in making the enclosure large; management's interest is in making it small (inexpensive). Who speaks for the maintenance man? He will at least have to go in to put sharpened blades in place. Once in a while he will need to pull a 50-hp motor and replace it. How does he do that? This is an especially good question if you understand his problem, management's requirement to spend less and make more, and your idea that the enclosure should be as large as possible.

Learn everything you can about how the operation is supposed to work *and supposed to be repaired*. There is a good reason to give the maintenance man room to swing a wrench inside that enclosure. It will speed up small repairs and adjustments. That ought to please management. Then there is a reason to design the enclosure so that people have room to work inside.

Now what about replacing that 50 hp motor? They weigh 500 lb. One great idea is to make a whole side removable. These are sometimes called "knock-out" panels. They can be pulled and the motor can be picked up and hauled away—and a new one brought the same way—with a fork truck.

You need to do some acoustical detailing so that there are no big cracks around these panels. If your bent is in structural engineering, you need to see that the structure is sound without support from the panels that can be knocked out at convenience. At least you should see that they will be easily sealed so that no cracks result from your detailing and that the panels themselves can be lifted, stored, replaced, and refastened without losing their acoustical virtue. This is a big order. If you are depending on a structural engineer, the problem is more complicated still.

As a first book in acoustics, this one takes all these ideas and questions very seriously. However, it does not guarantee that even memorizing what is said here gives any real answer, let alone the best answer, to such questions. If you are a past master at attaining good production rates, *and* a practicing structural engineer, and can appreciate the havoc brought about by "insignificant" changes in the acoustical design you are in good shape . . . and why in the world are you reading this book?

If you pretend only to know acoustics, look to your absorption, direct paths for noise radiation, mufflers (yet to come), seals, and of course TL through the enclosure walls. That is the basic minimum of what is expected of you. Now, try to put yourself in the position of management and of the production and maintenance people who will have to live with your creation. They have to make a living, too. If you cannot see what they—especially production and maintenance people—will be up against you are fooling yourself and your management. You had better believe that a fork truck is going to roll over that panel which is difficult to remove, and ruin it. Or that a fire will start inside the enclosure. Or that the machine can't be maintained. Or that a man lost his arm because he couldn't see what he was doing,

Neither I nor any noise control engineer I know has had someone lose an arm yet—but we have all vetoed designs that might have caused just that.

PROBLEMS

1 In problem 1 of the last chapter we found the TL of a *well-sealed* single wall of ⅝ in. gypsum board on studs. Those data are repeated below. Find the effective TL if the wall had had a ½ in. gap along one of its 8 ft long edges. Also find it for the case where a 1 ft² hole penetrated the wall. The total wall area is 64 ft².

Data and Answers

Frequency (one-third octave band)	Transmission Loss (dB)		
	Well-sealed Wall	With ½ in. × 8 ft Crack	With 1 ft² Hole
125	20	18	16
160	22	19	17
200	23	20	17
250	25	21	17
315	27	21	18
400	29	22	18
500	31	22	18
630	32	22	18
800	34	23	18
1000	35	23	18
1250	35	23	18
1600	35	23	18
2000	33	22	18
2500	26	21	17
3150	31	22	18
4000	37	23	18

2 *Worked problem.* Noise levels near a wood planer are about 113 dBA. Management proposes to build an enclosure, 16 × 32 × 12 ft high around the machine. It will be ½ in. plywood on 2 × 4 wood studs throughout. In the walls the studs will be on 24 in. centers. Since people will walk on horizontal surfaces, sooner or later, the top of the enclosure will have 2 × 4 wood studs on 12 in. centers. There will be two openings of 2 × 3 ft each to let the planks in and out of the enclosure. Can a 95 dBA level be expected near the enclosed planer? Near the openings? What effect will 3 in. of TIW (see Table 6) have on these levels?

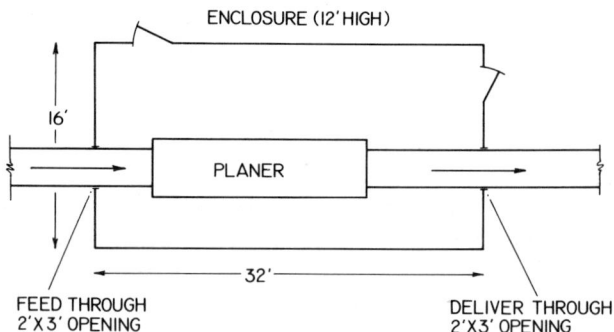

ENCLOSURE (12' HIGH)

16'

PLANER

32'

FEED THROUGH
2'X3' OPENING

DELIVER THROUGH
2'X3' OPENING

Method

(a) Since many important details are not given (ventilation, door and window information, for example) assume that this is an "order of magnitude" exercise to find out whether this type of construction has a chance of solving the problem.

(b) Estimate TL for walls and ceiling.

(c) Combine these TLs with those of the openings for feed and delivery.

(d) Estimate the reverberant build-up within the enclosure. From the combined TL found in step c estimate the level outside the enclosure.

(e) See what the use of thermal insulating wool will do to the reverberant buildup and adjust the figures found in step d.

Data

Space averaged levels near the planer are:

Octave Band (Hz)	Level (dB)	A-weight Correction (dB)	L_{eff} (dB)
125	90	−16	74
250	93	−9	84
500	99	−3	96
1000	108	0	108
2000	110	+1	111
4000	101	+1	102

—or 113 dBA.

The floor is painted concrete and about 20 ft^2 of floor area is typically covered to a depth of 3 to 6 in. with sawdust, small wood shavings, and chips.

Calculation

(a) To be order of magnitude estimate

(b) Find the TLs.

   (1) Find the wall and ceiling weights
   ½ in. ply weighs

$$\tfrac{1}{2} \text{ in.} \times 3 \text{ lb/in.} = 1.5 \text{ psf}$$

For 2 × 4 wood studs (1.5 × 3.5 in.) weighing 1.05 lb per running foot the
walls will weigh

$$1.5 + (1.05/2) = 2.0 \text{ psf}$$

and the ceiling will weigh

$$1.5 + 1.05 \doteq 2.5 \text{ psf}$$

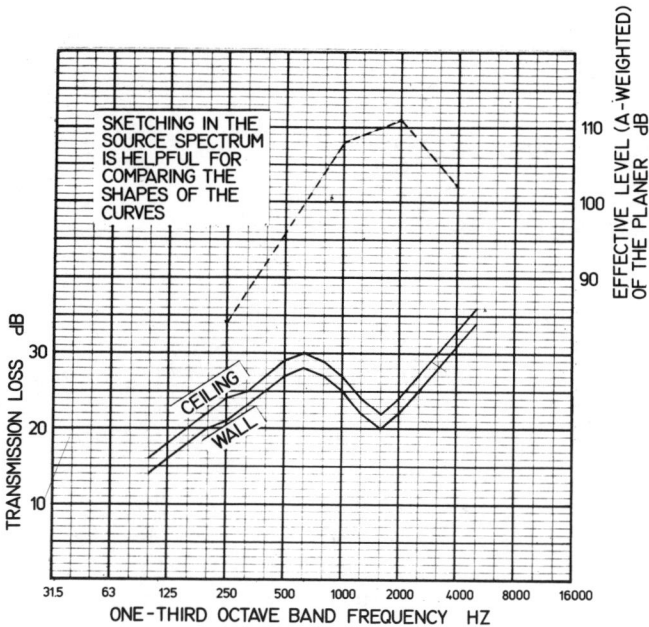

   (2) Develop the TL curves in the usual way.

(c) Combine the TLs

Octave Band	Wall TL	Ceiling TL	Openings
125	16	18	0
250	21	24	0
500	27	29	0
1000	22*	24*	0
2000	21	24	0
4000	31	33	0

Combine these using the nomograph (Figure 56) or make the calculation

$$\text{TL}_{\text{eff}} = -10 \log \left[\frac{S_{\text{wall}} \cdot 10^{-(\text{TL}_{\text{wall}}/10)} + S_{\text{ceiling}} \cdot 10^{-(\text{TL}_{\text{ceiling}}/10)} + S_{\text{open}} \cdot 10^0}{S_{\text{wall}} + S_{\text{ceiling}} + S_{\text{open}}} \right]$$

The areas will be

$$
\begin{aligned}
S_{\text{wall}} &= 12 \times 2(32 + 16) - 12* = 1140 \text{ ft}^2 \\
S_{\text{ceiling}} &= \quad\quad 16 \times 32 \quad\quad = \;\; 512 \text{ ft}^2 \\
*S_{\text{openings}} &= \quad\quad 2 \times 2 \times 3 \quad\quad = \quad 12 \text{ ft}^2 \\
&\quad\quad\quad\quad\quad\quad\quad\quad\quad\quad\quad 1664 \text{ ft}^2
\end{aligned}
$$

(Sample calculation: at 125 Hz

$$\text{TL}_{\text{eff}} = -10 \log \left[\frac{1140 \times 10^{-1.6} + 512 \times 10^{-1.8} + 12 \times 10^0}{1664} \right] = 15 \text{ dB})$$

Tabulating all the combined TLs, we have

Octave Band	TL$_{\text{eff}}$
125	15
250	19
500	20
1000	19
2000	19
4000	21

This will be the effective TL of the enclosure. Noise levels near the openings will be much higher, of course.

*Ordinarily you would read the curve where it intersected the octave band center frequency. In the bands indicated read the lowest $\frac{1}{3}$-octave band TL in that octave because the source and TL curves and diverging.

(**d**) Find the reverberant build-up in the enclosure.

(**1**) The planer before the enclosure

Item	Area (ft²)	125	250	500	1000	2000	4000
		Frequency					
Conc. floor (less 20 ft²)	492	0.01	0.01	0.01	0.02	0.02	0.02
		5	5	5	10	10	10
Sawdust (alpha for dry sand)	20	0.15	0.35	0.40	0.50	0.55	0.80
		3	7	8	10	11	16
Area walls will occupy	1140	1140	1140	1140	1140	1140	1140
Area ceiling will occupy	512	512	512	512	512	512	512
Area opening will occupy	12	12	12	12	12	12	12
Total sabins		1672	1676	1677	1684	1685	1690

(**2**) The planer after enclosure

Item	Area (ft²)	125	250	500	1000	2000	4000
		Frequency					
Floor		5	5	5	10	10	10
Sawdust		3	7	8	10	11	16
Walls	1140	0.3	0.25	0.2	0.17	0.15	0.10
		342	285	228	194	171	114
Ceiling	512	0.3	0.25	0.2	0.17	0.15	0.10
		154	128	102	87	77	51
Openings		12	12	12	12	12	12
Total sabins		516	437	355	313	281	203
ratio							
$\dfrac{\text{sabins before enclosure}}{\text{sabins after enclosure}}$		3.24	3.84	4.72	5.38	6.00	8.33
Noise buildup ($= 10 \log$ ratio in dB)		5	6	7	7	8	9
Original L_{eff} for planer		74	84	96	108	111	102
New L_{eff} inside enclosures		79	90	103	115	119	111

This means the level inside the enclosure is now 121 dBA.

(**3**) Now subtract the TL_{eff} found in step 3.

	125	250	500	1000	2000	4000
TL_{eff}	15	19	20	19	19	21
L_{eff} outside the enclosure	64	71	83	96	100	90

This means the level outside the enclosure is 102 dBA, so the 95 dBA target is not met.

(e) Reestimate the reverberant build-up if all wall and ceiling surfaces are covered with 3 in. TIW.

Item	Area (ft^2)	Frequency 125	250	500	1000	2000	4000
Floor		5	5	5	10	10	10
Sawdust		3	7	8	10	11	16
Walls	1140	**0.46**	**0.99**	**0.99**	**0.99**	**0.99**	**0.99**
		524	1129	1129	1129	1129	1129
Ceiling	512	**0.46**	**0.99**	**0.99**	**0.99**	**0.99**	**0.99**
		236	507	507	507	507	507
Openings		12	12	12	12	12	12
Total sabins		780	1660	1881	1668	1669	1674
Total sabins in unlined enclosure		516	437	355	313	281	203
ratio $\dfrac{\text{lined}}{\text{unlined}}$		1.51	3.80	4.68	5.33	5.94	8.25
NR = 10 log ratio (dB)		2	6	7	7	8	9
Level inside unlined enclosure		79	90	103	115	119	111
Level inside lined enclosure		77	84	96	108	111	102
TL$_{eff}$		15	19	20	19	19	21
Level outside lined enclosure		62	65	76	89	92	81

This gives a level of 94 dBA outside the enclosure. This is barely acceptable and the design ought to be carefully checked out. For example, levels will be much higher near the openings—they need mufflers. Also provision should be made to make sure no gappy seams or cracks will exist in the final construction. Above all, this calculation shows that the walls are skimpy. Leaks can be fixed, but it may be major surgery to improve the walls and ceiling later. A few more dollars can be well spent in making them better to start with.

3. *Worked problem.* A neighbor of an industrial plant has complained about a noisy fan on the roof of the plant. Measurements at the neighbor's bedroom window confirm the fact that noise reduction (see Data) is required to meet the local code. One proposal is to move the discharge duct to the opposite side of a roof monitor. Will this meet the requirements?

Data

Noise reduction required to meet local code

| | Noise |
Octave Band	Reduction (dB)
125	5
250	11
500	8
1000	3

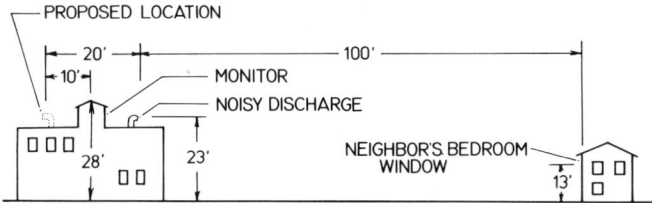

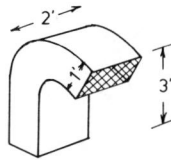

DISCHARGE DUCT OUTLET

Method

(a) Moving the discharge to the new location will reduce noise at the neighbor's house because of the increased distance. The noise reduction will be small, but it should be checked.

(b) The roof monitor will act as a barrier. Calculate its effect and add it to the NR found in step one.

(c) What about the effects of directivity? Could you simply turn the duct around to achieve the noise reduction? Is there any advantage in having it point away from the neighbor's house in the new location?

Calculation

(a) The effect of distance—the discharge is now at about 100 ft from the window and will be at about 120 ft in the new location. For these distances it can be considered a point source. For this situation,

$$NR_{distance} = 20 \log (d/d_o)$$
$$= 20 \log(120/100) \doteq 2 \text{ dB}$$

(b)   The effect of the monitor as a barrier:

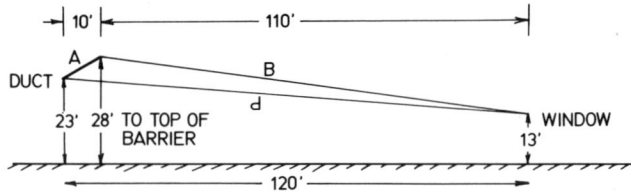

(1)   Work out A, B, and d from the geometry:

$A = \sqrt{10^2 + 5^2}$ $B = \sqrt{15^2 + 110^2}$ $d = \sqrt{120^2 + 10^2}$
$A = 11.18$ ft* $B = 111.02$ ft* $d = 120.42$ ft*

(2)   Now make the barrier calculation:
first

$$N = (F/565)(A + B - d)$$

then

$$NR = 20 \log\left[\frac{\sqrt{2\pi N}}{\tanh\sqrt{2\pi N}}\right] + 5 \text{ dB}$$

Octave Band (Hz)	N	$\sqrt{2\pi N}$	$\tanh\sqrt{2\pi N}$	NR Calculated (dB)	NR Required (dB)
125	0.394	1.573	0.9171	10	5
250	0.788	2.225	0.9769	12	11
500	1.575	3.146	0.9963	15	8
1000	3.150	4.449	0.9997	18	3

(The noise reduction predicted by the nomograph of Figure 57 will be the same as that calculated here.)

Thus the barrier effect will meet the NR requirements in every band. In addition, there will be at least a 1 dB improvement because the source has been moved farther from the neighbor's bedroom window.

(c)   What about directivity of the source? Since the requirements have been met, it need not be considered. However, suppose the solution had fallen a decibel or two short? Will redirecting the discharge help?

*All by the Pythagorean theorem. The unusual precision (for acoustics) is warranted here because you will be working with the *difference*.

From the shape of the discharge duct you might guess that it would be directive with a Q high enough to mean a few decibels difference between front and back. *But guessing can get you in trouble!* You might, however, make measurements around the discharge and see what the directivity effect actually amounts to. It will be greater as frequency increases, so measure in each octave.

By the same token if you find a marked effect do not be hasty by leaving the discharge where it is and directing it toward the roof monitor. This is a corner and will add its own directivity effect.

4. *Worked problem.* A control room overlooks a noisy chemical plant operation. The wall between them is brick but has a large window, two doors, and two ventilating air returns which allow the control room air to escape to the operation. The control room is noisy enough to be annoying and using the telephone is a little troublesome. Can the control room be made acceptable if the flanking paths through the air returns and cracks around the doors are blocked?

Data

Frequency		125	250	500	1000	2000	4000
Octave Band L_p(dB)		65	66	68	72	65	55

	Area (ft^2)		Transmission Loss, dB				
Brick wall	392	32	41	50	55	58	62
Window	160	20	25	31	36	36	30
Doors as hung, including cracks	42	8	12	13	15	14	16
Air returns	6	0	0	0	0	0	0
Comparable doors, gasketted	42	12	18	22	27	25	28

Method

(a) Assume all the noise is coming from the chemical operation.

(b) Find the existing TL_{eff} of the wall that separates the control room from this operation.

(c) Plan to move the air returns to another wall or duct them away so that noise cannot get into the control room by this path. Recalculate the TL_{eff}.

(d) Subtract the difference between the two effective TLs from the existing noise level.

(e) It will occur to you to find the A-weight level before and after the changes have been made. The difference in level before and after will give you an idea of the improvement. A better idea would be to find the PNC curve which applies in each case. This not only gives you an idea of the improvement but the curve can be compared to tables showing the suitability of the ambient noise to the space (Table 23 in the appendix).

Calculation

(a) Is all the noise coming from the chemical operation? Yes, all the noise that probably counts is—but you may want to make a trip to the control room and listen. Just possibly three people are usually talking loudly. Then you had better change your approach and look into absorption as the control method or as a supplement to this method.

(b) The existing TL_{eff} can be found by the nomograph (Figure 55) or the more rigorous calculation:

$$TL_{eff} = -10 \log\left[\frac{S_1 \cdot 10^{-(TL_1/10)} + S_2 \cdot 10^{-(TL_2/10)} + \ldots}{S_1 + S_2 + \ldots}\right]$$

(c) The improved TL_{eff} is found in the same way. When you have made both calculations a tabulation of your results should be:

Octave Band Frequency (Hz)	125	250	500	1000	2000	4000	A-weight
Improved wall TL (dB)	21	23	32	37	35	34	
Original wall TL (dB)	16	18	19	19	19	19	
(d) difference (dB)	5	5	13	18	16	15	
(e) The original level (dB)	65	66	68	72	65	55	**74 dBA**
will change to	60	61	55	54	49	40	**58 dBA**

You could note that this is an improvement of 16 dBA and the new room will be dramatically quieter than the other. But will it serve its purpose? Reference to Figure 9 will allow you to estimate that the room originally was at about PNC 70 or 75 and the improvement has it at about PNC 55. Referring to Table 23 in the Appendix you can see that PNC 70 or higher is not suitable for any work where attention and the ability to concentrate or communicate are important. PNC 55 is not ideal but it is acceptable in a large drafting room or steno pool. This may give you the idea that still more noise reduction, via absorption perhaps, might be appropriate in this space.

5. A fan discharge on a perfectly flat roof is exactly level with and 88 ft away from the bedroom window of a residence. The local noise code requires that no source of noise may produce more than 50 dBA at a bedroom window at night.

Since the fan must run all night, a proposal to build a barrier as shown in the sketch has been made. Will this meet the code requirement?

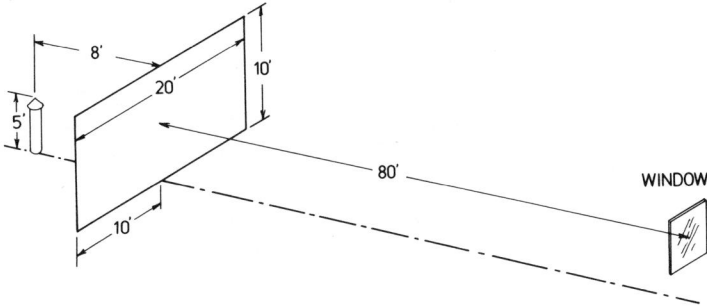

Data

Octave band data taken at the window on a clear, windless night when the fan was running show:

Octave Band	L_p (dB)
63	55
125	62
250	65
500	59
1000	51
2000	46
4000	42

Answer

If all the noise measured at the window came from the fan the barrier will just do the trick. (Remember to account for the sides as well as the top of the barrier.) Be assured that all the noise *did not* come from the fan. What is the answer then?

One good idea is to make the same measurements with the fan running and then with the fan off. Then you could say that you had controlled your source of noise adequately. Under some codes, the question could not be resolved at all.

6. Estimate the TL of a double wall of ⅝ in gypsum board on one side and ³/₁₆ in. plywood on the other. Framing is 2 × 4 wood studs 24 in. o.c. with no absorber in the cavity.

Answer

A 10 dB/octave slope was used in the double wall region and the arithmetic method was used to find the coincidence dip.

Band	TL	Band	TL	Band	TL
125	17	400	37	1600	52
160	19	500	40	2000	50
200	25	630	43	2500	43
250	30	800	47	3150	41
315	34	1000	49	4000	52
		1250	51		

10

MUFFLERS, SILENCERS, LINED DUCTS

The difference between mufflers, silencers, and lined ducts is that everybody knows what a lined duct is. It is a duct with sound absorptive lining. Nobody seems to be clear on mufflers and silencers. The terms probably shouldn't be used interchangeably, but they seem to be. To be sure, a lined straight section of duct, especially if it has "splitters," is a duct silencer. Mufflers can be the gadgets on your car, big diesel engines, or positive displacement blowers—but they can also be small attachments that screw into the air discharge vent of a pneumatic cylinder.

This family of devices is used to control the noise associated with moving streams of gas and sometimes to let process feed material, product, or scrap in or out of an enclosure without letting all the noise pass also. When the problem is a moving gas stream, three general problem areas are:

1 Some reciprocating mechanism (your car's engine, a lobed blower, or some other device, such as a ventilating fan) projects a regularly spaced stream of pulses into the air.

2 Small jets of high velocity gas—usually air but sometimes steam—driving into still air around them, shear the surrounding air and create an unstable interface. This boundary between the high velocity gas and the still air thrashes back and forth randomly, but at a rate in the audible range. This produces the easily recognized hiss of escaping steam or an air hose.

3 Sometimes a gas stream or jet is directed at an edge or across a cavity mouth to produce a tone, for example, high winds moaning in the telephone lines. When any sort of obstacle gets in the way of the jet, noise is produced. If the obstacle is a wire or a thin edge, the tones will be well defined. If the obstacle is bulky and irregular, a whole range of frequencies may be generated. If the stream is directed across a cavity mouth, an organ-pipe-like situation may also create tones—sometimes quite pure tones.

The most common example of regularly recurring sound pulses is probably the typical industrial fan. It has a scroll shaped housing and a fan wheel similar to a paddlewheel inside. Every time one of the tips of the blades passes the cutoff of the scroll, a puff of sound energy is generated. It will appear at the discharge and also at the intake.

Suppose you had a six-bladed fan with a shaft speed of 1160 rpm. The fundamental frequency this would produce is

$$\frac{6}{rev} \times \frac{1160 \text{ rev}}{min} \times \frac{min}{60 \text{ sec}} = 116 \text{ per second or } 116 \text{ Hz}$$

Harmonics of this frequency will usually also be prominent. You can expect to see peaks at 232, 348, 464 Hz and so forth, *usually* diminishing in intensity as the harmonic gets higher.

Lobed blowers and internal combustion engines produce a fundamental and string of harmonics in a similar fashion. If you will look at the way such devices work, you can perform an elementary analysis similar to the preceding one to relate the expected frequency peaks to the operating or shaft speed.

Obviously, you could design and build something, whether you called it a muffler or a silencer, to let the gas stream escape from these noise sources, but block the noise in the stream. Don't do it! "Rolling your own" here is not so good a description of what you are doing as reinventing the wheel! You cannot design and build such a device to be effective and have a low pressure drop as cheaply as you can buy one. The people who make their living selling these devices have the advantage of long experience, test results to guide them where special designs are needed, and all the fabrication and supply problems under control. They also have standard mounting hardware, standard joining methods, and a knowledge of the peripheral problems such as dust loading, pressure drop to be expected, and others you may not even foresee.

Yes, you may indeed have to design something to meet the needs of a special problem. Before you start on that exercise, do more than check the catalogs: make a few phone calls to reputable manufacturers of similar standard devices. Your problem has to be quite special to make it simpler to solve and cheaper to build than something they can furnish.

PRESSURE DROP

In fan systems where a lot of air is being moved and the power requirements are high enough to be interesting, pressure drop can be very important. In some systems, the increase in pressure drop caused by inserting a silencer can add several horsepower to the power requirements of the fan. The least odious result of that is a power bill that is too high. Perhaps the most is a process that doesn't work or at least doesn't work well, or at full capacity. The oddest consequence may be that you will create even more noise than before by starving the fan. Worse than that, the fan sometimes will not be able to find a stable operating

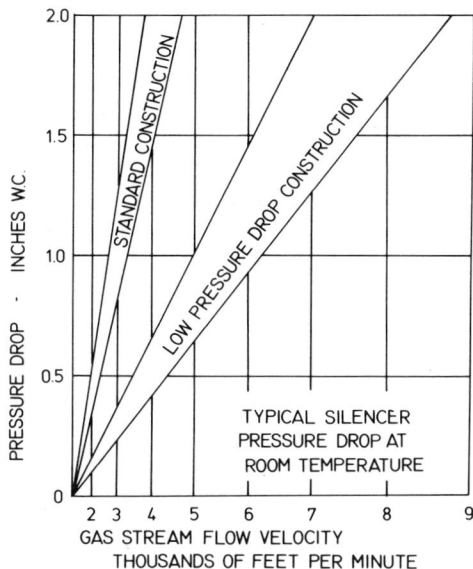

PRESSURE DROP - INCHES W.C.

STANDARD CONSTRUCTION

LOW PRESSURE DROP CONSTRUCTION

TYPICAL SILENCER
PRESSURE DROP AT
ROOM TEMPERATURE

GAS STREAM FLOW VELOCITY
THOUSANDS OF FEET PER MINUTE

Figure 59 These pressure drops are typical of commercially available silencers. Use them for rough planning only. When a specific silencer is chosen, work with the catalog data for it.

point on the fan curve and will (1) wear its bearings out with disgusting regularity and (2) sometimes even burn out motors.

For a variety of duct silencers, the sort of pressure drop you can expect to add to the system is shown in Figure 59. The data here are good enough for rough planning. The catalog data for the specific silencer you choose is *golden information*. In reading catalog information or that shown here, remember that it is based on good practice: a straight duct section out of the fan of five or at least three duct diameters so that local turbulences can be smoothed out and a tail section of duct at least three diameters after the silencer. If you are using the silencer as an intake, observe the recommended practices too—especially that of having a straight or at least *gently* curved inlet duct, if one is required.

REACTIVE MUFFLERS

Mufflers and silencers, as names, may lead to some confusion. However, there is a useful way of describing some devices of this sort. Some of them, like the duct silencer mentioned previously, make use of absorption and are called *absorptive* silencers or mufflers. Sometimes they are also called dissipative devices. The point is that the sound power is used up in the same way it is in conventional absorbers.

A typical design and a typical performance curve for an absorptive silencer (or muffler) are shown in Figure 60. The dimensions of the device are intentionally omitted and you should note the general shape of the performance curve rather than its particular values. In general, the shape will be like this for absorptive silencers. The frequency at which the attenuation peaks will be governed by the physical dimensions, in part, but also by the choice of the absorbing material, its thickness, and other details. If you needed 15 dB of attenuation at 232 Hz, this design would do it—and might be a reasonable and economical choice despite the "surplus" attenuation at the higher frequencies. On the other hand, if you had several of these fans to equip, you should search for the most economical design—or even have a standard design modified.

Another kind of noise control device—like the muffler on the tailpipe of your car—uses no absorptive material at all and still does a good job. This is a *reactive* muffler. Reactive mufflers make use of tuned pipes and cavities in ways that are modifications of the Helmholtz resonator.

The simplest scheme for applying this principle is shown in Figure 61. Notice that this lets gas flow down the straight run of pipe with no obstruction in the path. The arrangement is called a side-branch resonator.

What you are doing is to introduce into the duct, at the side branch, ·a pressure front *just when the normal flow calls for a rarefaction front* (and, conversely, rarefaction and pressure front, of course). This is a clever idea if the sound wave moving down the duct is a sine wave—everything cancels out.

Now, in fact, even better performance is attainable by leading the gas stream in at one end of the resonator volume and letting it escape somewhere else, usually at the other end. The effects attainable in this way are shown in Figure 62.

Once again, we are not going to bother with how these reactive mufflers are designed because it does get tricky and it will never be realistic for you to try to design and build one as cheaply as you can buy it. There are excellent units

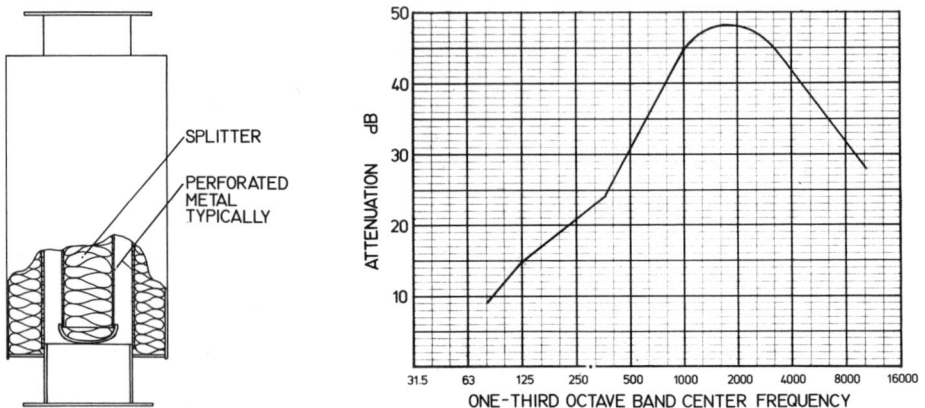

Figure 60   A section through and performance curve of an absorptive silencer. In a rectangular duct, the center piece will be a splitter; in a round duct, a cylindrical core.

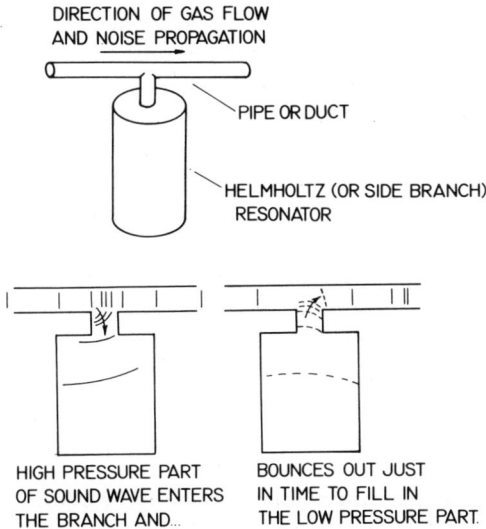

Figure 61 The side-branch resonator works on the Helmholtz idea. As sound waves pass down the pipe, wavefronts enter the cavity and bounce out again just in time to fill in the rarefaction part of the sound wave.

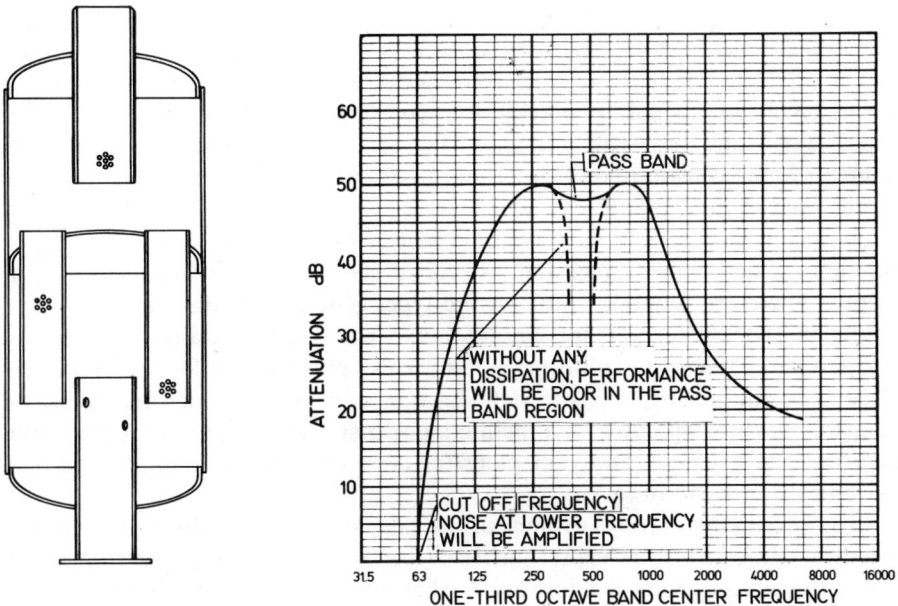

Figure 62 Construction and performance of a reactive muffler. No dimensions are given. The performance curve is specific to the size and design. Note that the connecting stub tubes are perforated. This furnishes "dissipation" (partly detuning, but also some dynamic loss) that prevents the performance from sagging badly in the pass band.

available for industrial use—some of them approaching a railroad tank car in size. The published data are fairly good. Do observe the following caution for both silencer and muffler performance curves.

The manufacturer will have taken pains to make an ideal installation when he tested the unit or had it tested by an independent lab. In a typical installation in your plant, there may not be room for nice smooth transition sections into and out of the unit. You may have to come into the muffler from an elbow, for example. It is good practice to allow a 5 dB margin between what you need and what the curve says you will get. Allow more if you can see that you'll have to make a really bad installation of it. Without smooth transitions, the pressure drops also increase greatly from the published data.

JETS

The last of these items that you cannot design and build as cheaply as you can buy them are mufflers for air exhausts. High pressure plant air gets exhausted for one of two reasons, usually:

1 To move or direct the motion of something.
2 Because it has done its job in moving a piston and must be cleared out to make way for the return stroke.

To quiet the first of these, there are a variety of designs that preserve the thrust (in fact, one enhances it) while reducing noise. They are sketched in Figure 63. Even in small quantities, they cost less than $10 each ($3 to $6 is the usual range) and they will pay back their cost in a matter of months! They reduce the air consumption, and compressed air is one of the most expensive things used in most plants.

In the case of exhaust air, two basic designs exist. In one, the air has to pass through a plug of fiber (steel wool or bronze wool) or a porous surface (sintered metal). These are superior to the second type, generally, in the amount of noise reduction they achieve. Plant air, however, most often contains oil and these mufflers are prone to plugging. The back pressure builds up until the pneumatic equipment begins to slow down or even stops. Somewhere along the line, the machine operator or a maintenance man figures out what's happening and fixes the machine (and your plans for noise control) by taking off the muffler and throwing it away.

To prevent this, always explain to the people concerned about the plugging problem. Have plenty of spare mufflers (they're cheap enough) and even tape one to the equipment or exhaust line so that it is on hand when a replacement is needed. Also, make sure that the maintenance department knows that soaking a plugged muffler for a day in solvent will make it usable again. There are devices for taking the oil out of the air. The problem is that they fill up or plug

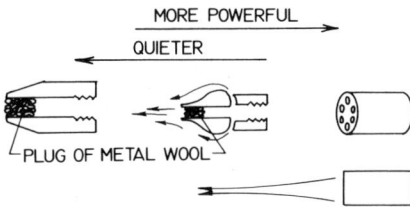

Figure 63   A few of the designs for nozzles and thrust mufflers used to quiet air jets. Note the streamlined shape of the center design. Surrounding the jet with a shroud of air moving in the same direction minimizes shear between the jet and the still air surrounding it. In the design on the right several small holes are arranged so that their jets will converge (Bernoulli's principle). This arrangement is much quieter than an open pipe or hose. Surprisingly, although it uses less air than an open hose, it produces a greater thrust.

MORE POWERFUL

QUIETER

PLUG OF METAL WOOL

up themselves. Refrigeration units will drop out both oil and water, but they are expensive to buy and increase operating costs too. They can be justified only in special cases.

The other sort of muffler can still provide as much noise reduction as you usually need. It is a small cylinder ($3 \times 1\frac{1}{4}$ in. diameter and sometimes much smaller) with a series of slots in the side. The air escapes through enough openings to be diffused and the volume of the cylinder acts as a reservoir—someplace for the incoming blast to accumulate.

A BASIC IDEA

Jets make noise by shearing the surrounding (still) air. The interface between the high velocity jet and the surrounding air is unstable. The interface shifts back and forth at audio frequencies and produces the noise. The interface can be thought of as a hard surface radiating the noise. It is not, of course, and its efficiency as a noise radiator is poor, which is fortunate for noise control people, too, because the interface has a lot of power behind it.

If you could do nothing else, you might consider increasing the size of the exhaust line—preferably in a series of stages from $\frac{1}{4}$ in. pipe to $\frac{1}{2}$, 1, 2, and 4 in. and so forth. By allowing the gas in the jet to expand, you are reducing its velocity relative to the still air. For a simple jet, the sound power generated varies as the cube of that velocity. If the jet must encounter an edge, the sound power varies as the fifth power of the velocity.

There should be a long enough run of pipe after each increase in diameter for the flow to stabilize in the pipe. If you had your choice, you would put smooth transitions between the diameters. Probably you'll reject this approach on every job you face. It is a little odd. The idea behind it is not! Where and when the elements of the idea can be fitted into some other requirement easily, make use of it.

STEAM

When you get to steam, you have a new set of problems. It's hard to reduce the velocity of steam because it keeps expanding (at the expense of its temperature) and will defeat a good many of the mufflers that work beautifully for air. Then there's the most obvious thing about steam: it condenses and will turn nice new hardware into rusty junk. There are special mufflers for steam. They are expensive. In view of cost, it is usually better to pipe it away to someplace where the noise won't bother anybody. Be sure to use traps to allow the condensate to clear out of any low points in your piping.

Piping the steam away is fine for small steam vents. What do you do in the situation where a paper machine drier goes down and you need to dump the steam being generated by a 50,000 lb/hr boiler? The general features of a steam blow-off muffler are shown in Figure 64. It combines some features of reactive and absorptive mufflers (the fuzzy section in the center represents mineral wool packed between the perforated tubes leading to the top vents). What happens first as the steam enters the bottom is that it is given the chance to expand and shed some of that deadly velocity. Performance of such mufflers depend on steam rates and pressures, of course, as well as on the design details. For the frequencies above 1000 Hz—generally the most intense for blowing steam—some designs have a fairly flat performance curve of better than 50 dB.

LINED DUCTS

Let's go back to the muffler section we talked about for getting planks in and out of the enclosed planer. The section is simply a piece of duct or a box with the two ends missing. It blocks some of the noise because the walls, top, and bottom have been lined with an absorber. The same arrangement can sometimes ac-

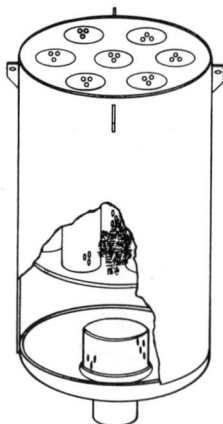

Figure 64 Steam blow-off mufflers can handle rates of tens of thousands of pounds of steam per hour.

complish enough to keep fan noise from being annoying at the ventilation register.

Usually you will not need to predict performance of one of these devices because catalog data is abundantly available. Sometimes, though, you will be lining a duct or building a lined tunnel muffler (the planer in-feed and delivery, for example) and will need to estimate performance.

It may be disheartening to learn that the four methods commonly used to make such predictions are often in strong disagreement with each other. Forecasting what will happen for ducts is not as tidy as predicting TL of a wall and it is still less reliable than predicting the effect of absorptive surfacing in a space. However, although none of the methods take prizes for accuracy, they share a common virtue: they are generally conservative. You should do better than you predict.

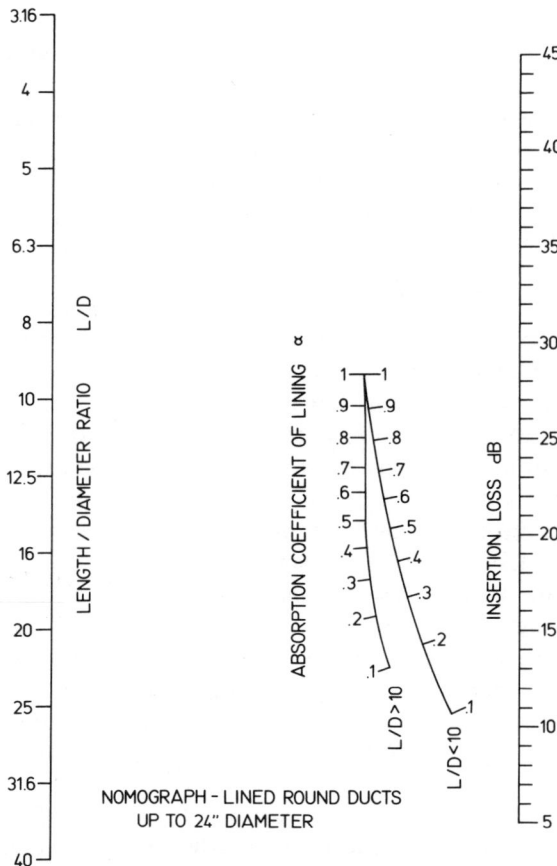

Figure 65 This nomograph approximates the insertion loss for a lined round duct when you know the absorption coefficient of the lining. For long ducts use the center scale on the left; for short ones, the scale on the right.

It is not worth much of your time to look into reasons for this sloppiness in prediction. Still, there may be an insight or two lurking in the question. First, how can it be that using absorption to quiet a space is handled with fair reliability, but using the same material in a duct is uncertain? One reason is the forgiving nature of the NR $= 10 \log(A_f/A_s)$ relationship. It was based on averaging the absorption over the whole possible range of angles of incidence for a flat sample in a reverberation room. In a duct, unless it is huge, the dimensions of the section are small in terms of wavelength. A whole raft of effects hinging on angle of incidence complicates the situation.

Second, there may be some attenuation by reflection of the sound from internal section changes or even the opening through which the gas stream issues. This gain in performance from reflection is usually a low frequency phenomenon.

Third, the same sort of reflective complication can prevent the noise from getting into a duct efficiently. Don't count on that happening with fan noise getting into the discharge duct. It may be a help at the return register of a room or the tunnel inlet and delivery to a large enclosure around a planer.

It is paradoxical that where the duct section is huge, so that absorption might have a chance to work predictably, the problem is usually that the duct surfaces won't allow enough absorption to be brought to bear. The usual solution to this, while effective, takes you right back to a restricted range of angles of incidence. You insert absorption lined baffles in your duct so that they run parallel to the direction of flow. You may have to increase the physical dimensions of this part of the duct in order to accommodate the absorptive baffles without reducing the sectional area of the duct too much. In fact, if the velocity is high, you will probably want to protect the fibrous absorber with perforated metal. There are many other details of construction, like streamlining, you might want to take into account, but this is exactly what you must avoid.

If you need a lined duct and modest effort in design and construction will fill your needs—go to it. When you begin to roll up a whole exercise in streamlining, protecting the absorber, increasing the section, and then *building* the duct, rest assured that the experts in designing and building them can meet your requirements at far less cost than you can.

The estimation methods also assume that the lined section is part of a continuous run of duct. Where that is the problem you're solving, the estimation methods are more reliable. However, if you are designing a tunnel muffler where both ends abruptly terminate in a large space—one at a wall, perhaps, and the other projecting into the space of a larger room—your estimate of performance may be way off. Reflective effects take place at terminations like these.

Two of the methods are not given here at all. The first of these—a complex impedance calculation using Smith charts—has the best reputation. It is a tedious, drawn-out method, however, and is somewhat beyond the scope of workaday practical requirements. If you decide to employ the method anyway, beware of a scale error of a factor of 10 in the impedance ratio scale (it should read 1 to 1000, not 0.1 to 100 as widely published). The other method omitted here is ray tracing to estimate how much acoustical power can survive a trip through the lined portion. It is equally time-consuming and is a method of desperation used when the geometry is much wilder than that of a lined duct.

ESTIMATING PERFORMANCE

For quite small ducts, there is a very simple method indeed. It works so long as the width and height of the duct section (to the inside of the lining) are less than a tenth of the wavelength of the sound.

For this case:

$$A_d = 12.6\alpha^{1.4} \, (P/S)$$

where A_d = the attenuation of the duct (dB per running foot of duct)
α = the absorption coefficient of the lining for the frequency of interest (no units)
P = the perimeter length of the duct section (in.)
S = the sectional area of the duct (in.^{2})

The difficulty lies in the size of duct for which this works.

Frequency (Hz)	Wavelength in Air at Room Temperature (ft)	Maximum Duct Dimensions (in.)
63	18	22
125	9	11
250	4.5	5.5

You can see at a glance that this simple approach is not often applicable.

For usual section sizes and frequencies, the design chart of Figure 68 will be required. It predicts that at some low frequency, there will be little effect from the lining and that, as the frequency rises, the attenuation will rise to a peak and then decline again.

It is safest to believe what Figure 68 predicts. You will often be pleasantly surprised to find that things are not as bad as predicted at the higher frequencies. The smaller the duct dimensions, the more true this is. When you consider what is happening in the physical system, you can see why. If the duct dimensions are several feet by several feet and the length is not very great, the short waves of the higher frequencies can pass through headlong without being absorbed. Catalog data and your own measurements for small ducts with reasonably high length to section ratios show better performance than predicted for high frequencies.

To use the design chart, you must first come to grips with two of the prime variables: resistivity and d/λ.

Resistivity (see, the old horse traders of Chapter 6 *did* know what they were talking about) is usually fairly difficult to track down. It has the reputation of varying significantly from lot to lot of "identical" material and even from spot to spot in the same board or batt. You will very rarely find it in the manufacturer's literature and asking the salesman will probably not produce it either.

Table 10   *Resistivity Data for Glass Fiber Board (resistivity in mks rayls/meter)*

Type of Board	Typical	Range
1.5 pcf (Owens-Corning 701)	10,000[1]	7,600–14,000
3 pcf (Owens-Corning 703)	24,000	17,000–35,000
6 pcf (Owens-Corning 705)	31,000	24,000–40,000

[1] In other units, 10,000 mks rayls/meter would be 26 cgs rayls/inch or 260 SI units. (Data courtesy of Owens-Corning.)

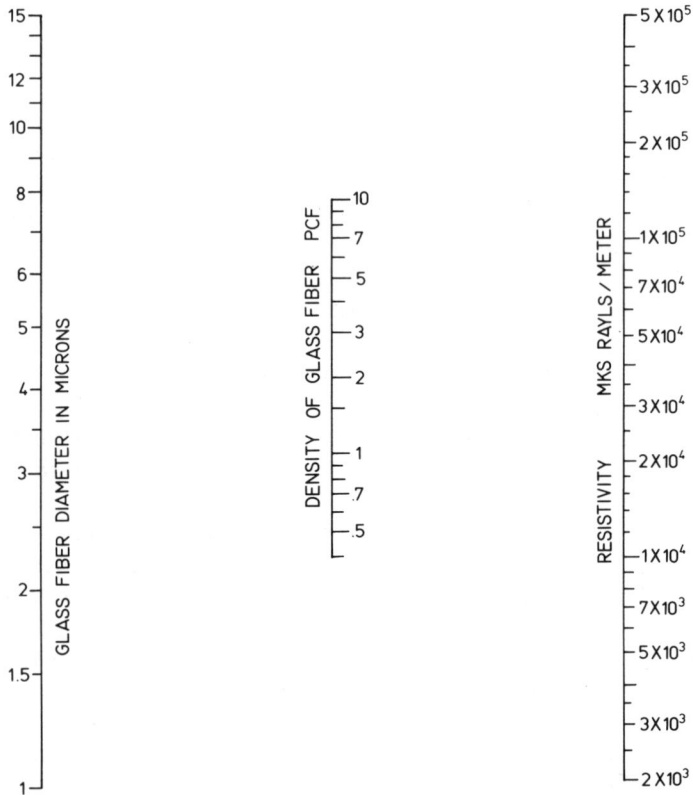

Figure 66   If you can find the fiber diameter of glass fiber batt or board, this nomograph will produce a usable figure for the resistivity in mks rayls/meter. (400 mks rayls/meter = 1 cgs rayl/inch.)

You might try phoning for technical assistance, but on this question even the most helpful technical people don't like to commit themselves. Now, if by chance you do find a number for resistivity, be sure to check the units. The figure you want is in mks rayls/meter. If you are given a number in cgs rayls/inch, just multiply it by 400. (See Table 10.)

If you cannot get a figure for resistivity, you probably *can* get the average fiber diameter of the glass fibers that make up the lining. In that case, the nomograph of Figure 66 will produce a useful indication of the resistivity. The big problem with the nomograph is that it cannot handle lining materials that are not homogeneous. Moreover, some of the best liners—especially for high velocity air—have either a felted or neoprene latex reinforced surface.

The second of these problem quantities—d/λ—will give you no trouble once you understand the rules. The wavelength λ is familiar and can be found easily as soon as you know the temperature of the air. If you should be working with some other gas, you will need to find the wavelength using the method in Chapter 1. The tricky quantity here is the dimension d. The calculation method always assumes that two opposing surfaces of the duct are covered. If only one of that pair is, the duct will behave as though it were twice as large in that

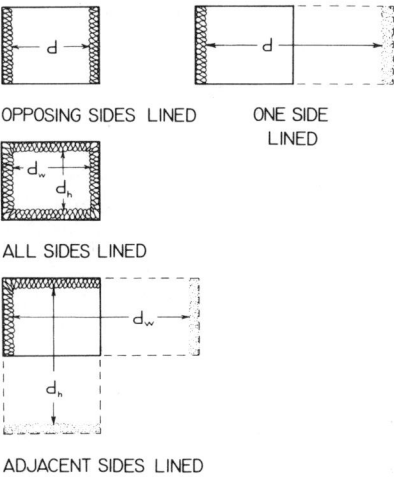

OPPOSING SIDES LINED   ONE SIDE LINED

ALL SIDES LINED

ADJACENT SIDES LINED

Figure 67 In using the design chart (Figure 68) you need to work with a duct dimension d. Where opposing inner faces are lined d is the distance between the inner surfaces of the lining in feet. Where only one of the opposing faces is lined d is the distance from the liner to its reflected image. When two adjacent or three or all four sides are lined the calculation is made for each pair.

dimension. Also, if surfaces in both dimensions (the height and width of the section) are lined, you will need to run the calculation twice, adding the attenuations found. Figure 67 serves as a quick reminder of how to treat d. Do remember to keep d and λ in the same units, whether inches, feet, meters, or for that matter cubits!

Figure 68 looks forbidding and it has one or two tricks up its sleeve that *are* annoying the first time or two you use it. There is a reason for this. This design chart brings together an enormous amount of data in a small space. The chart

EXAMPLE

d = 3

P = .75

125 HZ

A = 1.6 dB/d
OR ~.5 dB/FOOT
SINCE d = 3 FEET

8
6
5 — 60
4 — 48
3 — 36
2 — 24
18
1 — 12
9
6
3

FEET
INCHES

DUCT DIMENSION

d

FREQUENCY HZ

63
125
250
500
1000
2000
4000
8000

DO NOT USE THESE
SCALES UNLESS YOU
ARE WORKING WITH AIR
AT ROOM TEMPERATURE

125   2   3.15   5   8   1.25   2   3.15
1   16   2.5   4   6.3   1   1.6   2.5

.0063
.008
.01
.0125
.016
.02
.025
.0315
.04
.05
.063
.08
.1
.125
.16
.2
.25
.315
.4
.5
.63
.8
1.
1.25
1.6
2
2.5
3.15
4
5

λ/d

15 OR MORE
10
6
4.5
3.0
2.25
1.75
1.25
1.0
.85
.75
.65

RESISTIVITY
PARAMETER

P

1   16   2.5   4   6.3   1   1.6   2.5
125   2   3.15   5   8   1.25   2   3.15

ATTENUATION IN dB/d
(WHERE d IS THE DUCT DIMENSION)

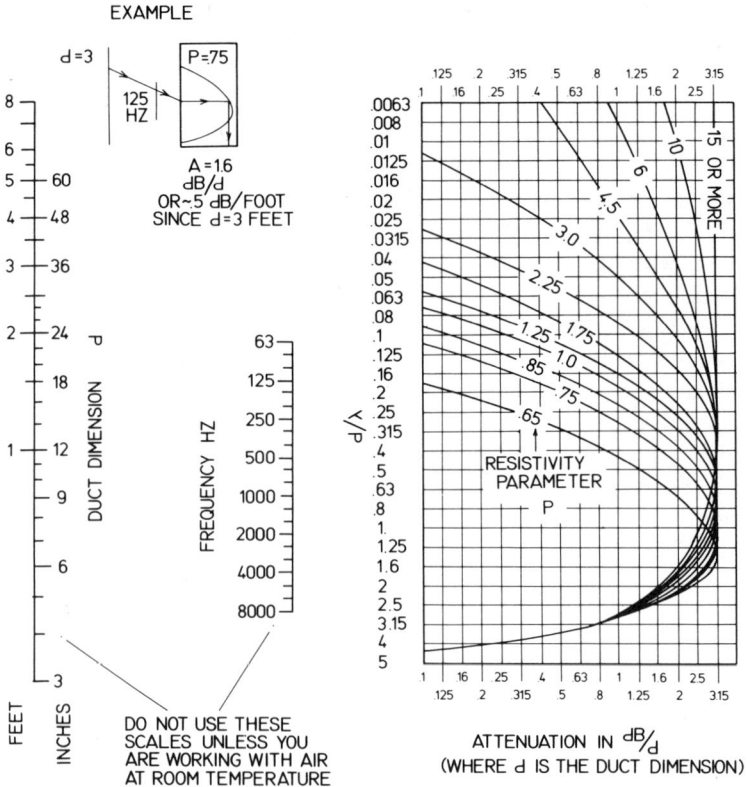

Figure 68 Design chart for lined ducts. If you are working with any gas stream other than air near room temperature do not use the nomograph scales at left. Alignment of duct dimension d and frequency will locate d/λ on the chart (or you can calculate λ to find d/λ for hot or cold air or other gases). Run across the chart at d/λ until you intersect the proper curve for parameter P. Then travel vertically to the top or bottom attenuation scale to find how many dB in a length of duct equal to dimension d.

could be made much simpler—and published, in modified form, again and again as one of the four key variables was increased while the others were held constant. It would be a book as big as this one! You've already cleared the air on a number of points that give trouble the first time out. Smile now and pick your way through Figure 68. One day you will look on it as an old friend.

NOW LET'S HAVE A POSITIVE ATTITUDE

First you need to get that resistivity data into a number which the chart recognizes. To do this, find the parameter P:

$$P = Rt/16,000$$

where P = a resistivity parameter (no units)
        R = resistivity (mks rayls/meter)
        t = lining thickness (in.)
       16,000 = a constant with units that harmonize mks rayls with inches

Now, if you are working with air at room temperature, put a pencil point down on the proper d of the left duct dimension scale and pivot a straightedge on this point so that it also rests on the frequency you are interested in on the center scale. The straightedge will intersect the d/λ scale. Trace a horizontal line from the intersection until you come to the curve of the proper resistivity parameter. Then drop vertically to the bottom scale of the graph. For gases other than hot or cold air, you'll have to find λ (Chapter 1) and enter the d/λ scale of the plot directly.

The attenuation is given in decibels per length of lined duct equal to the duct dimension d. Therefore, divide the lined length by d and multiply the attenuation by that factor (the number of d's contained in the lined length). If the duct has a lined surface on the adjoining face or faces, you must repeat the calculation for that pair too.

WORKED PROBLEM

A ventilating fan causes a level of 87 dBA near the registers. The spectrum is given below. The air is supplied through a 24 × 30 in. duct. One 10 ft run of straight duct is easily accessible and can be lined on all four sides with some 2 in. 1 pcf material having a resistivity of 18,000 mks rayls/meter. What effect will this have on the level near the registers?

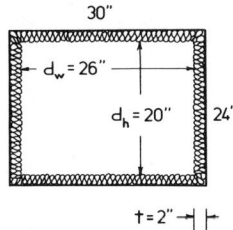

30"

$d_w = 26"$

$d_h = 20"$ 24"

$t = 2"$

Data and Initial Calculation of L_{eff}

Octave Band Frequency (Hz)	Original L_p (dB)	A-Weight Correction (dB)	L_{eff} (dB)
125	91	−16	75
250	90	−6	81
500	86	−3	83
1000	80	0	80
2000	72	+1	73
4000	68	+1	69
			or 87 dBA

Calculations

The first thing to do is find the resistivity parameter P.

$$P = Rt/16{,}000 = (18{,}000 \times 2)/16{,}000 = 2.25$$

Second, find d and then the attenuation for the longer side of the duct. The duct dimension d will be 26 in. (see sketch). Enter the nomograph with 26 in. and 125 Hz. You will read about 0.25 as d/λ and be able to trace over to an attenuation of about 2.8 dB in length d. Since 26 in. is 2.17 ft, there are 4.6 d's in the 10 ft lined section and the total attenuation for this pair of faces will be 2.8 $\times$ 4.6 = 13 dB.

Third, by the same method find the total attentuation for the other pair of faces at this frequency.

Finally, add the total attenuation for both pairs of faces in each octave and subtract this from the L_{eff} for that octave. Find the new A-weighted level in the usual way.

Your results should look like this:

Octave	Attenuation (dB) for d = 26 in.		Attenuation (dB) for d = 20 in.		Total Attenu- ation (dB)	L_{eff} (dB)	
	In 26 in.	Total	In 20 in.	(dB)		Old	New
125	2.8	13	2.5	15	28	75	47
250	3.15	14	3.15	19	33	81	48
500	3.1	14	3.15	19	33	83	50
1000	1.8	8	2.5	15	23	80	57
2000	0.2	1	0.8	5	6	73	67
4000	—	—	—	—	0	69	69
			Calculated A-weighted level			87	71

Thus the lined section has produced a reduction of 16 dBA.

The design curves we have just used (Figure 68) suppose that the inner surface of the liner is covered with perforated metal. The fraction of the metal covering which is open should be matched to the resistivity parameter (P). The preferred materials are:

Resistivity Parameter	Percent Open	Resistivity Parameter	Percent Open
0.75	80	3	33
1	73	4	22
1.5	60	5	18
2	50	>5	15
2.5	40		

The effect of not using the metal covering is that the low frequency performance will not be as good as predicted. Nonetheless, these design curves are often used to predict lined duct performance even when perforated metal is not used. One reason for this casual attitude is that believable resistivity data are hard to find. Another factor is that it usually doesn't take much absorber to accumulate a big predicted attenuation. The calculation is used to see that the material and dimensions have matched the frequencies that need to be attenuated, then an ample run of lining is prescribed. For muffler sections on in-feeds and deliveries from enclosed equipment, it is a good idea to use the perforated metal anyway, simply as a matter of protecting the absorber from moving objects.

LINED ELBOWS

The most effective place to use lining in a duct is at an elbow. Here, the noise has to collide head-on with the absorbing surface. The usual recommendation is to line the duct for at least four times the maximum duct dimension on both sides of the elbow. It is better practice to go six times the maximum dimension. If this has been done, you can use the preceding calculation method to find the basic attenuation and then come back and put in a bonus for the elbow. For frequencies whose wavelengths are less than one-third of the larger duct dimension, add 10 dB to the attentuation you've calculated. For all frequencies with wavelengths two-thirds or more of the major duct dimensions, take no added credit. In the region between, the curve will rise from 0 to 10 dB.

Using the duct from our worked problem as an example, the larger dimension is 26 in., or 2.17 feet. A third of this 0.72 feet. That wavelength is associated with 1130 fps/0.72 ft = 1600 Hz. The frequency with twice that wavelength is 800 Hz, of course. If we had used enough lining in the previous problem to have satisfied the "six times the major duct dimension" requirement, we could claim the bonus performance sketched in Figure 69 by simply moving the lined portion to an elbow.

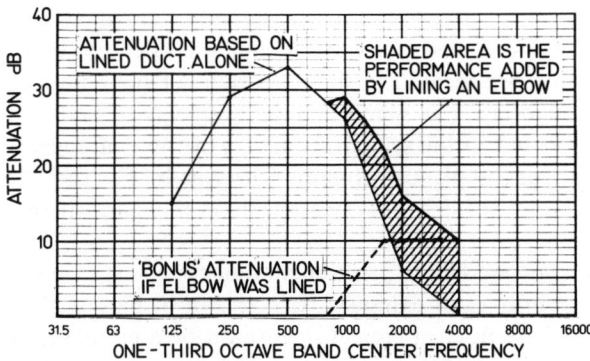

Figure 69 Placing the duct lining at an elbow in the duct produces an increase in attenuation for some higher frequencies.

Be a little wary of calculations that predict attenuations of 60 dB and more. Flanking is a factor in mufflers, too. For one thing, noise will escape through duct walls because they are usually light sheet metal with limited TL. For another, even if you succeed in blocking passage of sound through the air stream, the walls of the duct will convey vibration past the lined section and reradiate it downstream of the lining. Finally, where the air velocity is high, noise will be generated in the air stream by turbulence.

We are not going to get into ventilating and air conditioning here—we've more than enough to do in our bailiwick. However, because you *are* concerned with noise, do what you can to have good practice employed in air moving systems. Use good transitions, turning vanes, wide sweep or mitered elbows where possible, and easy versus hard turns. The greater the power of the system, the more important it becomes.

PLENUM CHAMBERS

Sometimes you find that a big fan with a spectrum rich in low frequencies can't be controlled with anything you can think of. Lined ducts call for materials that don't exist, or calculated pressure drops for conventional mufflers are too high. This is where you reluctantly turn to the big gun of the muffler family, the plenum chamber. It is expensive, but sometimes it's the only thing that will work.

A plenum chamber is sketched in Figure 70. Typical sizes begin at or about the size of a single car garage. The calculation method given by the formula looks forbidding, but you'll find it easy to use once you unscramble all the unknowns. Note that neither frequency nor wavelength are mentioned. They are accounted for by the absorption coefficient. As long as you know alpha for the frequency you're interested in, you can calculate the attenuation. This calculation also is troublesome when you have frequency peaks below the usually reported range for absorbers.

Plenum chambers are sometimes used end-to-end. A large chamber is constructed and an internal baffle turns it, in effect, into two such chambers.

Notice that the inlet and outlet of the chamber are shown as slots. That is not a trivial detail. You need not worry about the openings you use as long as they are slotlike and are located in simple corners of the chamber. The money-saving change that often gets made in your design is to make them square or round openings located in the trihedral corners of the chamber. Look out! This can alter the effect of the plenum chamber drastically. For some frequencies the performance not only suffers, it is possible that the performance will drop to zero.

A junior sized version of the plenum chamber is a surprisingly effective muffler for ventilating intakes or exhausts. It is shown in Figure 71.

For a 12 in. square duct, a typical size would be a cube 3 ft on an edge or a little larger. This is one muffler you can undertake to build. Plywood and glass fiber board are typical materials.

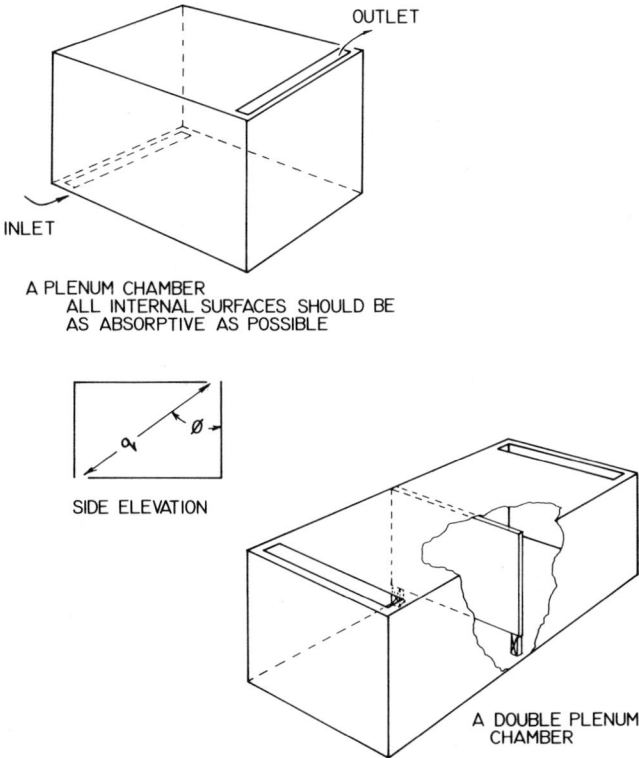

A PLENUM CHAMBER
    ALL INTERNAL SURFACES SHOULD BE
    AS ABSORPTIVE AS POSSIBLE

SIDE ELEVATION

A DOUBLE PLENUM
    CHAMBER

Figure 70   The plenum chamber muffler is sometimes used in pairs, end-to-end. For any single section find the attenuation using

$$A = 10 \log \left[S \left(\frac{\cos \Phi}{2 \pi q^2} + \frac{1 - \bar{\alpha}}{a} \right) \right]$$

where S = the area of the outlet opening (ft²)
   Φ = the vertical angle of the shortest path between inlet and outlet
   q = the shortest path length, inlet to outlet (ft)
   $\bar{\alpha}$ = the average absorption coefficient of all surfaces for the frequency
   a = the total absorption of the inner surface (sabins)

The attenuation you find will be a negative number. You can change its sign before subtracting it from the original level or simply add it to the original level.

PROBLEMS

1   Worked problem. A small transformer vault needs to be ventilated. Convection will move the air and no fan is required. If you use a round duct 12 in. in diameter with 1 in. TIW as lining, how long a run would you need at the inlet to achieve a 6 dB reduction? At the discharge to get 10 dB? The only important noise is a pure tone at 240 Hz.

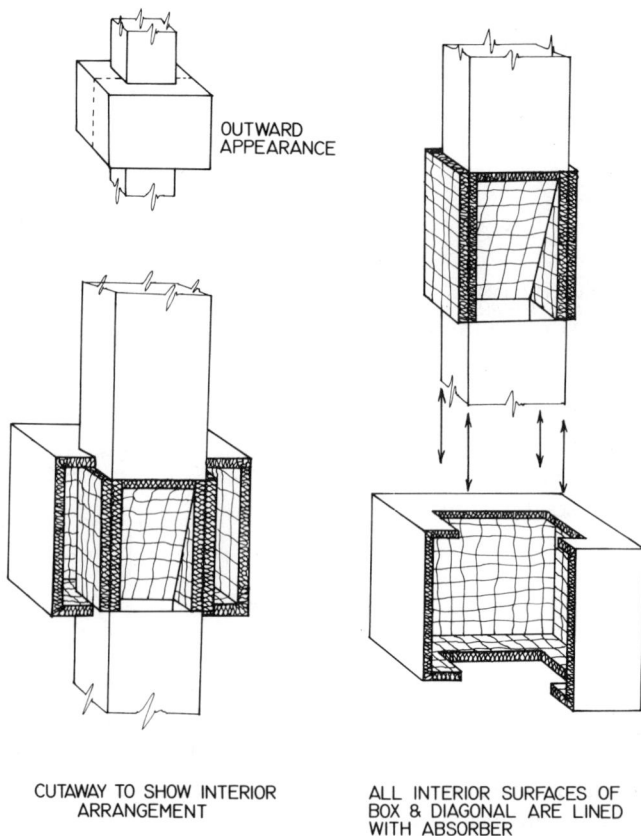

OUTWARD
APPEARANCE

CUTAWAY TO SHOW INTERIOR   ALL INTERIOR SURFACES OF
ARRANGEMENT   BOX & DIAGONAL ARE LINED
                                  WITH ABSORBER

Figure 71 A clever variation on the plenum chamber design is sometimes used in ventilating. It is not recommended for high velocity air, but it is a high attenuation silencer. Start by cutting openings in two opposite sides of the duct. Then block the duct off with a diagonal so that when it is enclosed in a box the air will exit through one opening and reenter through the other. No simple method exists for estimating performance, but it will be the highest possible per cubic foot of silencer.

Method

(a) It is tempting to use the $A_d = 12.6\ \alpha^{1.4}\ (P/S)$ equation. A check of the wavelength shows that this will not work. Instead, use Figure 65.

(b) To do this, you need an alpha for TIW at 250 Hz. It is given in Table 6.

(c) Enter the nomograph with the required NR on the right-hand scale. Align this point with the alpha and read how many diameters the run must be.

CALCULATIONS

(a,b) Table 6 gives 0.33 as the absorption coefficient for TIW at 250 Hz. That alpha is for a number 4 mount—not much like the duct.

(c)   Enter the nomograph at 6 dB and align an alpha of 0.33. Your straightedge runs off scale but shows that about 2.5 diameters would be required. Since the lined diameter is 10 in, this will be 25 in. or 2 ft. By the same method, estimate that you will require about 4 ft for the 10 dB discharge.

Comment   Not what you would call an exact solution because of the question about the value of alpha! Still, the requirements ought to be met without great expense even if you are very conservative in calling for longer runs of duct, even twice what you calculated. Be assured, too, that the conservative nature of the predictive method is working for you here.

2   Worked problem. Openings of 2 ft high × 3 ft wide have been left at the in-feed and delivery to a planer. These are the principal leaks in the enclosure and you would like as much noise reduction at these two locations as you can get. You must have at least 6 dBA reduction. Space is limited and the tunnel mufflers you plan to use may not be longer than 8 ft. The planks will be carried on a chain conveyor which is 24 in. wide and 11 in. deep (from the top open surface of the chain to the bottom of the returning section). Using 4 in. mineral wool and a covering of perforated metal about 15 to 20% open, can you find a muffler design that will achieve 6 dBA?

Data

Frequency	L_{eff} Inside the Enclosure (dB)	
125	77	Mineral wool of 6 pcf density has a re-
		sistivity of 100,000 mks rayls/meter
250	84	Initially you submit a design sketch
500	96	(sketch 2) showing all four sides of the
1000	108	tunnel lined.
2000	111	Management suggests that more clear-
4000	102	ance at the top and an unlined floor
	or 113 dBA	are preferred (sketch 1).

Method

(a)   Evaluate both schemes (sketch 1 and sketch 2).

(b)   Find d_{width} and d_{height} first.

(c)   Find the resistivity parameter P.

(d)   Use the design curves, Figure 68, to find the attenuation for an 8 ft length of duct for both width and height.

(e)   Add the total attenuations for width and height.

(f)   Subtract this NR from the L_{eff} by octaves and then find the new A-weighted level. Compare it with 113 dBA for the existing openings.

Calculations

(a) The general features of the two schemes are shown in 1 and 2.

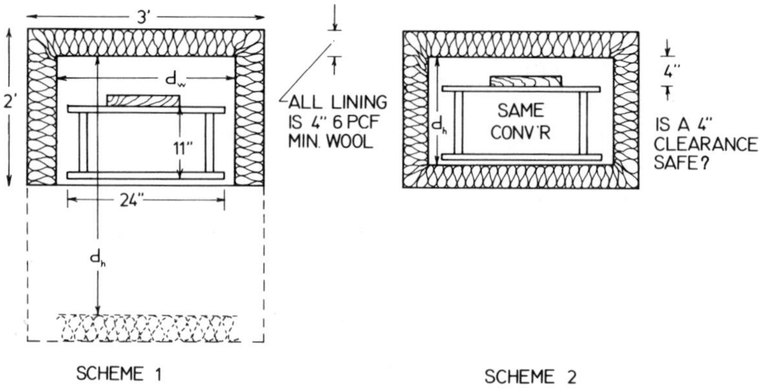

SCHEME 1   SCHEME 2

(b) Whatever wall material is chosen for the tunnel mufflers, suppose that they can be built so the inside dimensions of the unlined tunnel are 2 × 3 ft. Then the critical dimensions of the muffler section are:

	Scheme 1	Scheme 2
d_{width}	28 in. or 2.33 ft	Same
d_{height}	40 in. or 3.33 ft	16 in. or 1.33 ft

You might also question the height dimension because of the chain conveyor and planks. Be assured that the chain is mostly open space. When the planks have passed through, there will be times when the planer is making its loudest noise. The effects of planks and chain can only be better than your estimate of peformance by reducing the effective duct dimension d. It is better to plan for the worst case.

(c) $P = (100,000 \times 4)/16,000 = 25$

(d) Performance of the 8 ft long muffler section:

	Scheme 1 (28 × 40 in.) Attenuation (dB)					Scheme 2 (28 × 16 in.) Attenuation (dB)				
Octave	In 28 in.	Total	In 40 in.	Total	Sum (dB)	In 28 in.	Total	In 16 in.	Total	Sum (dB)
125	3.15	11	3.15	8	19	3.15	11	3.15	19	30
250	3.1	11	2.8	7	18	3.1	11	3.15	19	30
500	2.7	9	1.8	4	13	2.7	9	3.1	19	28
1000	1.4	5	.7	3	8	1.4	5	2.3	14	19
2000	0.1	0	0	0	0	0.1	0	1.25	8	8
4000	0	0	0	0	0	0	0	0	0	0

(f) Subtracting the attenuation, by octave from the L_{eff}, produces a new A-weighted level of 112 dBA for scheme 1 (net gain 1 dBA) and 106 dBA for scheme 2 (gain of 7 dBA and thus meeting the requirement). Both designs will probably do better than predicted but scheme 1 may not meet the requirements. The problem exists with the higher frequencies and better absorbers will not help much—though a more open perforated metal covering would be some help. The obvious need is for a longer run and smaller section.

3 If you lined an elbow in a square duct so that the inside dimension was 6 in. what Noise Reduction would you expect beyond that available from a straight run?

Answer

No improvement below 3400 Hz and 10 dB above 6800 Hz. (Actually, it will do better than that.)

11

VIBRATION ISOLATION

Vibration isolation has to do with breaking a path for sound energy in the solid members of some structure. Fairly obviously, if the only way that energy can get to the operator is through a structural path, then breaking that path is an effective control measure.

Let's start with a simple example. You may have had a house built, you may be building one, or, perhaps, you dream of building one some day. One of the annoying things about a house is that when somebody flushes a toilet, it is heard in every room that is quiet. It must have been airborne noise or it wouldn't have gotten to the listener's ears ... but how? Usually not directly through the air. The usual way that the sound of a flushing toilet gets to every room in the house is shown in the source-path-receiver diagram of Figure 72.

The important path is through the water to the pipe wall, from the pipe wall to the wood framing of the house, from the framing to the large, light panels (floors and walls) of the house, and thence from them to the ear. The pipes are hung from the wood structure. Now for very little cost, you can make it very difficult for structure-borne sound to jump from the pipe to the structure. For example, you can specify that oversize clips be used with a wrapping (between the pipe and the clip) of oxhair felt, mineral wool, or, as a reasonable material, old cotton towelling or strips from old blankets.

Can you calculate what improvement you can expect from this technique? There are methods of estimating the noise reduction that such a scheme would produce. They are hardly worth the effort in this case. Your intuition should tell you that there is a lot to be gained and neither the material nor labor cost will amount to enough to worry about in any new construction. If you can anticipate a noise problem and see a solution this easy, just use it.

SIMPLE INDUSTRIAL PROBLEMS

As we go on, you will see that breaking the structural path is sometimes costly. If you call for this technique and do not block such a trivial path as bringing the electric power in by means of coiled flexible (armored) cable, you will be letting the vibrational cat out of your expensive bag. The problem is far worse if you

SOURCE   PATH   SECONDARY SOURCE   RECEIVERS

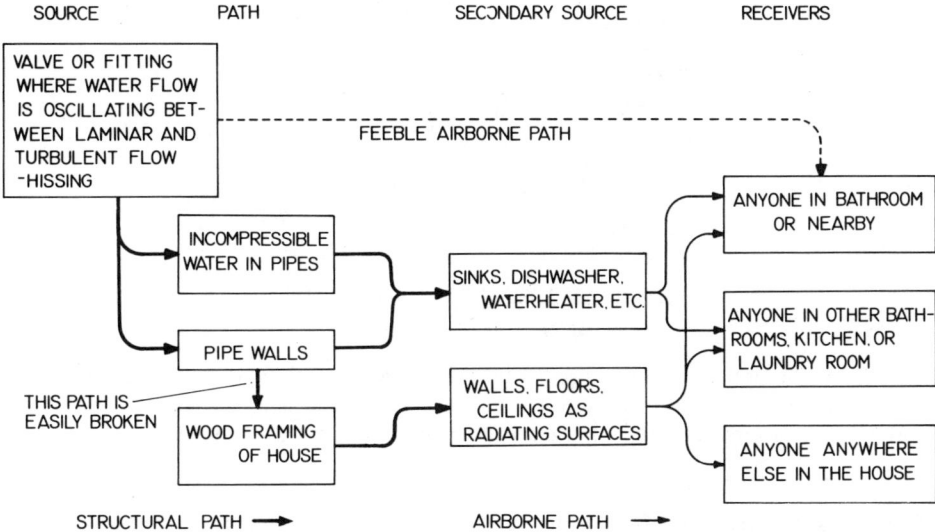

Figure 72   The sound of a flushing toilet or running water is often heard plainly in every room of a house. For those rooms away from any plumbing, breaking the structural path at the lower left gets rid of this nuisance.

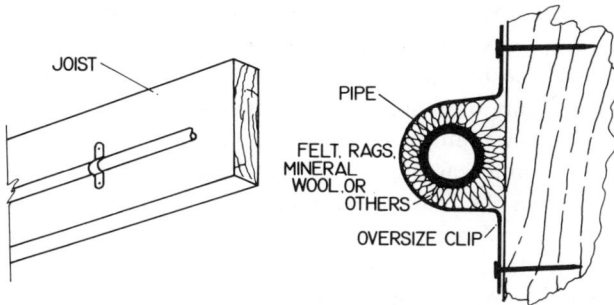

Figure 73   This simple vibration break is an inexpensive way of breaking the critical structural path shown in Figure 72.

have to connect the vibrating equipment to city water or a high pressure oil line. You cannot keep such connections from being rigid and hence structural paths for vibration. Even the use of rubber hose (which can be dangerous if the pressure is high enough) doesn't solve the problem because the high pressure incompressible fluid acts as a vibration path. The scheme shown in Figure 74 is sometimes used. The oil line is suspended from spring isolators and a heavy weight is hung on it. (If you have a background in electrical engineering or electronics, and if you'll turn the arrangement upside down in your mind and change the symbols appropriately, you'll recognize a π-section filter. This, of

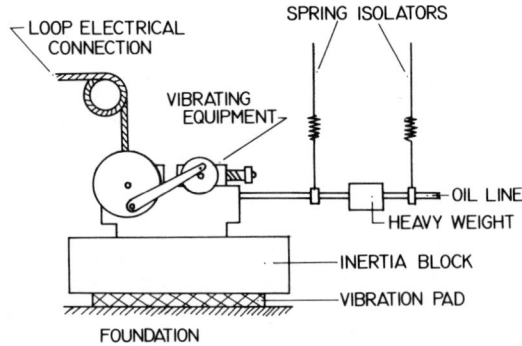

Figure 74 Several commonly used vibration isolation techniques are used in this hypothetical machine setup. In addition to the vibration pad and inertia block, isolators and weights on pipes and flexible electrical connections can be important.

course, is exactly what it is.) Be careful in using this scheme. Strengths and especially the length of pipe must be chosen to accommodate the vibration without breaking.

The main feature of Figure 74 is the fact that the vibrating equipment has been attached to an inertia block and mounted on a vibration pad. The pad is resilient. (If you recognized the π-section filter described previously, this time you can see that the arrangement works like a choke-input filter.)

Here's what happens. The equipment tries to vibrate at some frequency (or several). We are supposing that it is trying to hop up and down. (Actually, it is trying to move in many directions.) The inertia block was chosen to be utterly rigid—concrete heavily reinforced with steel, for example—and weighs at least four or five times as much as the equipment. If floor loadings will take it, an inertia block of 10 times the mass of the equipment is often used, especially if the equipment tends to vibrate strongly. The energy causing the vibration is "diluted" by having to move the additional mass.

The pad, or sometimes springs, accepts the vertical vibration and its top surface moves to conform with the travel of the block. At the bottom surface, if everything has been properly designed, the force is greatly reduced. It is fairly easy to reduce the force by a factor of 10. Factors of 100 can be attained. In rare cases, even greater reductions are possible with big equipment. In theory, almost any reduction could be had but the vibrations in other than the vertical direction usually impose limits to these great reductions.

CHOOSING AN ISOLATOR (PAD OR SPRING)

In order to avoid being swamped with mathematics, it is usual to work with what is called a "single degree of freedom" system. The mass and spring shown in Figure 75 will help clear up the preceding comments about the many ways the

SINGLE DEGREE OF
FREEDOM THEORY
ACCOUNTS FOR THIS

MASS

SPRING

BUT ALL OF THESE
(AND MORE)

CAN HAPPEN, TOO

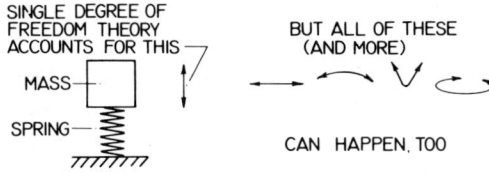

Figure 75 The single degree of freedom assumption concentrates on the most common—and usually most important—mode of vibration. It is prudent to remember that even simple vibrating systems may be vibrating in more ways than that assumption describes.

system tries to vibrate. In the simple theory, we suppose it only tries to move vertically. Actually, it is also trying to move in the other ways shown, too (and more, besides).

For this system and considering the pure vertical vibration only, there is a characteristic behavior curve (Figure 76). The peak of the curve is where the driving frequency of the vibrating equipment equals the natural frequency of the system.

To predict this natural frequency, you can use the relationship

$$f_n = 3.13 \sqrt{1/\delta}$$

where f_n = the natural frequency of the system (Hz)
  δ = the deflection of the isolator under the static load imposed by the mass (in.)

Consider first what will happen if the equipment wants to vibrate at the natural frequency of the system. Every time the equipment starts upward, the isolator and inertia block (if one was used) are also urging it upward. When it starts down, they want it to come down too. The amplitude in such a resonant system will build up and the results can be catastrophic. If there were no damping (that word again) in the system, theory tells us the amplitude of vibration could ap-

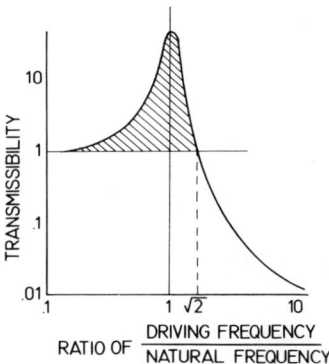

Figure 76 The general behavior of an isolated vibrating system considers transmissibility—the ratio of dynamic force exerted on the foundation before and after isolation. Note that when the driving frequency equals the natural frequency of the system the transmitted force is much greater.

proach infinity. Every real system, however, has something in it that drains some of the vibrational energy out.

Steel springs provide the least damping of the common isolators. Here, about ½% of the energy is wasted by "internal friction" in the springs. Natural rubber or neoprene soak up about 5%, felt or cork about 6%, and special materials and systems will take as much as a third of the energy out of the system.

This is important when you are designing a mount for a big diesel or any big reciprocating equipment that must start slowly and work up to speed. It will have to pass through the natural frequency of the system. You may even have to provide restraint (removable clamps) for slow starting equipment. You will at least want a fairly well damped isolator.

Politics may be the *art* of compromise, but engineering is the *science* of compromise. You will pay, in performance, for the damping in the system when the equipment gets to its normal operating speed. A family of curves for the same natural frequency is sketched in Figure 77. Damping certainly limits the worst

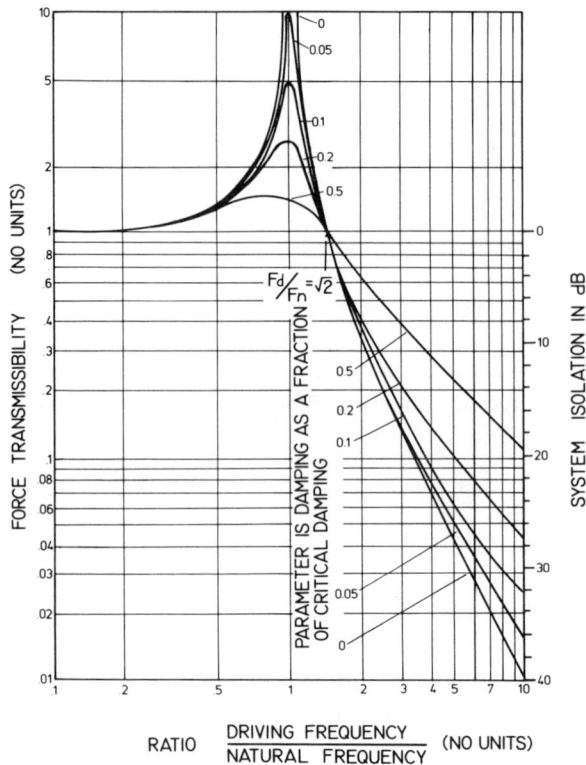

Figure 77 The vibration isolation design curves. Optimists may read the system isolation in decibels directly from the right-hand scale.

case. It also reduces the isolation you can achieve. That isolation is usually called transmissibility, and the scale is logarithmic. The numbers tell you what fraction of the force gets through the isolator.

If you wanted to, you could use 20 times the log of the reciprocal transmissibility. This would be an insertion loss for the isolating system in decibels. This is *not* a very handy approach, though, because while the loss through the isolating system may be well known, it is not usually possible to say what will happen once the vibrational energy gets into the supporting structure. You must also remember that the single degree of freedom assumption is hiding other vibration modes and their effects. Forces imposed by them will usually exceed the force of the isolated mode at some frequencies.

To the right of the natural frequency peak the curve swings down, crossing the transmissibility of 1 at 1.414 times the natural frequency. Not only will you want to stay out of the range below $f_d/f_n = \sqrt{2}$, you will want to design your system for driving frequencies of 3 to 10 times the natural frequency.

THE HARDWARE

Steel springs are much used for heavy equipment or for collections of equipment mounted on a single inertia block. The three facts that you should know about steel springs, as distinct from other isolators, are that

1 They work splendidly for the lower frequencies but, depending on their dimensions, they begin to "leak" vibration at higher frequencies. These high frequencies don't see the "springiness" of the spring but spiral their way around the coil and get into the structure.

2 In order to reach the required deflection, it is sometimes tempting to use a height-to-diameter ratio that leads to instability. The springs will buckle to the side.

3 Because fairly long springs are needed to get the required deflection, there is often a clearance space of several inches under the inertia block. If the system doesn't provide the isolation you think it should, check this space to be sure it *is* clear. Trash or scrap seems to find its way under the block by magic. Even a half-crushed soft drink can will transmit an amazing amount of vibration.

At the other extreme from steel springs are the many sheet materials manufactured for this use. They include natural rubber and neoprene, often reinforced with cotton or other fiber, specially bonded glass fiber board (about 18 pcf), cork, felt, and so forth. For high loading and extreme durability, long fiber asbestos felt is sometimes sealed in a lead envelope as a vibration pad.

A hotel built in 1914 rested on these pads (it was just above the main tracks of the New York Central) and when it was torn down to make way for the Union Carbide Building in the late 1950s, some of the pads were still in good enough

Figure 78 One of a wide variety of rubber pads used for vibration isolation.

shape to be used again. The sealed lead casing has also been used with rubber pads when they were to be installed in a pit and exposed to oil or water.

Avoid cork. It has been used as an isolator for years and it works very well when it is fresh. Oxidation slowly hardens it. A cork pad in service for 15 years is often rock hard, and of course, it doesn't work anymore.

In addition to springs and pads, there are a host of patent isolators and, by and large, they are the best isolators available. Often they have an adjustment for increasing or decreasing the damping in the system. They often incorporate leveling screws so that the machine can be trued up or the load on various isolators equalized.

ESTIMATING PERFORMANCE

The transmissibility curve has an x axis that is the ratio of driving frequency to natural frequency. For a system like rubber or neoprene pads, the damping will be 0.05. (We said these materials would drain out about 5% of the energy. In fact, that is quite an approximate statement. The 0.05 is the ratio of damping supplied by the material to the critical damping of the system—the amount of damping that just prevents free vibration.)

For a system with rubber pads and a natural frequency of 10 Hz, the following performance can be (optimistically) expected:

Driving Frequency (Hz)	Transmissibility T (no units)	Reduction in Vibration $[dB = 20 \log (1/T)]$
20	0.45	6
40	0.075	22
60	0.035	30
80	0.022	34
100	0.016	36

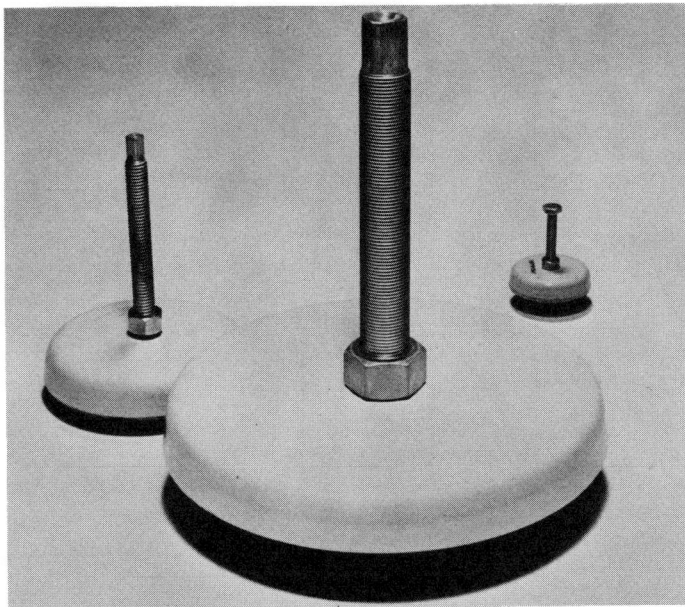

Figure 79 A common type of patent vibration isolator. Its resilient material is neoprene and it incorporates leveling. (Courtesy of Barry Controls)

The estimates, especially in decibels, must be looked on as approximate. The performance may be much better than predicted at some frequencies. On the other hand, all the vibrational modes not included in the single degree of freedom approach do exist and some of them may appear in amplitudes higher than you estimate at some frequencies.

Suppose you were troubled by vibration from a motor-generator set weighing 800 lb. The set turns at 1800 rpm. What could you expect if you put two layers of rubber padding under each of the four corners of the machine?

Figure 81 gives deflection data for two $^3/_8$ in. rubber pads stacked on top of each other.

Let's suppose we cut the pads into 2 in. squares. Four of these stacks would give a bearing surface of 16 in.^{2}. Dividing the 800 lb load by 16 produces a loading on the pads of 50 psi. From the deflection data, we can see that δ will be 0.15 in. Plugging this into the equation

$$f_n = 3.13 \sqrt{1/\delta}$$

produces $f_n = 8.08$.

The driving frequency of the set is 1800 rpm divided by 60 sec/min or 30 Hz. The ratio of driving to natural frequency is, therefore, 30/8 = 3.75. Check back to the transmissibility curve (Fig. 77) for a damping of 0.05 and read a transmissibility of about 0.08 at that ratio. In decibels, this would be 20 log (1/0.08) = 22 dB.

Better isolators for this job exist, as was pointed out earlier.

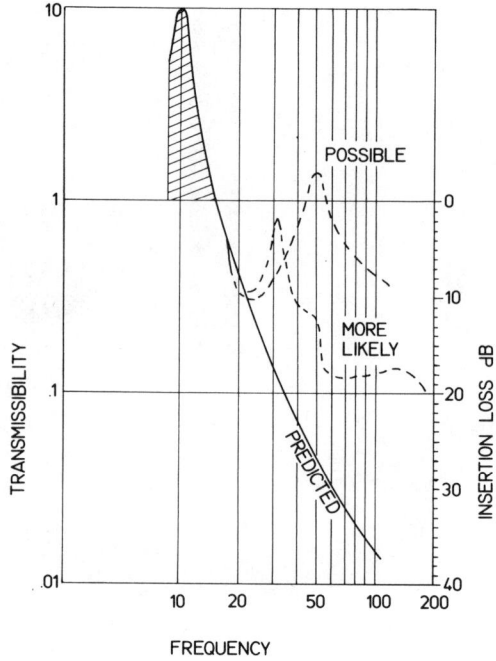

Figure 80 Theoretical prediction and some other possibilities of the isolation of a system when the natural frequency is 10 Hz.

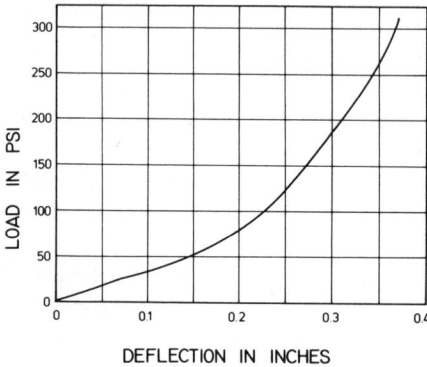

Figure 81 The deflection curve for two $3/8$ in. rubber pads stacked on top of one another. This is typical of available catalog data.

PATENT ISOLATORS

For loading from tens of pounds to tons, there are patent isolators that offer very good performance and the convenience of leveling and load adjusting. They are based on one of three resilient members: steel springs, rubber, or air.

In some of the steel spring designs, there is a wrench adjustment separate from the leveling screw that will increase or decrease the damping. Damping can be raised to as much as 30% in some designs. This will not often be required. One place where it might be is under a hammermill or jaw crusher where, in addition to steady vibration, you can anticipate severe shock loading from time to time.

Rubber is the workhorse isolator material and there are more patent isolators based on rubber of one kind or another than are based on the other materials. In general, there are two ways in which the rubber is employed. It is either compressed by the load or placed in shear by it. Some rubber mounts combine both methods. For the same amount of material or size of the isolator, you can usually expect better performance but lower permissible loading for the rubber in shear designs. Straight compressive designs accept higher loads and are often less expensive though they don't perform quite as well.

Just because there are so many variations available in rubber mounts, it is a good idea to get in touch with the technical service department of isolator suppliers when you are choosing these mounts. For one thing, they usually have had experience with the specific job you are doing and can direct you to the two or three designs and sizes that are the most economical and best. For another, they can steer you past pitfalls due to torsional loading and a host of other practical problems.

The most critical work is usually done with air isolators. They would be the choice if you had to install an optical bench in the same building with heavy equipment. It is commonplace to design the system to have a natural frequency as low as 1 Hz when some types of air isolators are used. At the same time, air isolators are an excellent choice under such brute force equipment as punch presses. They do well here because the very low natural frequency can furnish some isolation even for the low frequency components of the force exerted by the press.

GREMLINS AND GLITCHES

Unless you have a great deal of experience behind you, you should seek advice as you work on a vibration isolation project. The reputable manufacturers of these devices all maintain offices that will give you technical help without

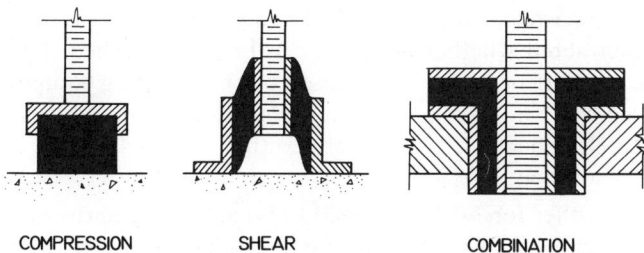

COMPRESSION   SHEAR   COMBINATION

Figure 82 Three ways of applying the load to isolators based on rubber.

Figure 83   Air isolators are high performance devices, they are available in sizes from 25 lbs to 10 tons per mount. (Courtesy Barry Controls)

charge. Some of the mistakes they can either steer you away from or at least warn you about are:

1   The theory behind the design curve and prediction assumes that the foundation is utterly rigid. If you are isolating one component on a machine, you will probably be close enough to that condition for the method to be useful *if* the supporting member of the machine weighs 10 or 20 times as much as the component and is quite stiff. Similarly, the prediction works well for equipment isolated from bedrock and well enough for most equipment on a slab at grade. However, if you try to isolate the engine of a portable air compressor from its frame and housing of about equal weight—or a punch press from the middle of a span of the second floor of the building—there is very little you can accomplish. A fairly complicated calculation (the technical service department will help you with it) can give you an idea of whether any reduction is possible. In the two bad cases, here, about 5 dB would be an optimistic forecast. A rule of thumb is sometimes used—at least in coming to a decision about whether anything can be accomplished. Find the static deflection of the floor in the span caused by the vibrating equipment. (Don't be surprised by $1/2$ in.!) To be effective, your isolator should have six or eight times as much deflection. If an isolator with $\delta = 3$ to 4 in. can be used, you may have a solution.

2   Consider the other forces! If you need to isolate a long narrow diesel engine, the most natural idea would be to put the isolators directly under it. After all, you need to support the weight that will be directly above them. Did you

Figure 84 Air isolators are a favorite for sources like punch presses because of the low frequency content of the vibration.

ever watch a diesel several hundred hours after the last overhaul wake up on a cold morning? There is a great deal of coughing and hit and miss running until it warms up. This produces some side to side swaying and jerking, and *that* won't do your overloaded isolators any good at all! Always check to see whether you need to broaden the support base or arrange for some (isolated) lateral snubbing.

3   Besides flanking your isolators, all the connected services (electric power, oil, air, gas, water) will be coming from someplace that is not vibrating (we hope) and connecting to something that is. You need to get them to where you want them without ruining the isolation, as has been pointed out. You should consider also that the pipe or hose or cable through which they flow may be subject to severe stress. A broken gas line near a hot exhaust header is something nobody needs!

4   Obviously, you should check the compatibility of materials. A spill of light hydrocarbons can raise havoc with natural rubber.

FLOATING FLOORS

In looking at enclosures, it was noted that sometimes it is not possible to enclose a noisy operation or the equipment making the noise. In that case, you can often design a booth to enclose the operator. In order to save money, you will use light construction for the booth. A light structure of this sort is particularly prone to flanking by vibration in the supporting floor. The light walls and ceiling will pick up the vibration. You can see the trouble you are in then.

It is possible, of course, to build a nice rigid frame, support it with any of the isolators described previously, and construct the booth on that. In fact, that is a very good idea if you are building a cab type enclosure fastened to the steel structure of a headsaw or a burr debarker. More often, you will be putting up your booth on a dry concrete floor. A simpler alternative here is to use blocks of 18 pcf glass fiber under the floor framing members. Depending on floor loading, they will be spaced at 12 to 24 in. centers both ways. An even more convenient way to float a floor is with a glass fiber blanket with the 18 pcf blocks imbedded in it. This is cut to size and a sheet of $^{3}/_{4}$ in. ply laid on top as the floor. The rest of the construction is built on the ply.

There is no tidy way to estimate the insertion loss achieved by this floating construction—but then there is no way of accurately forecasting what noise level would be induced in the unisolated booth by vibration either. The isolation will be very good at the higher frequencies that count most in determining the A-weight. It is certainly safe to expect that a floating floor of this sort is good for 20 or 30 dBA of isolation for any usual structural vibration.

GASKETS

If your job encompasses all the requirements of meeting OSHA regulations, you will know all about mechanical guarding to keep hands out of gears and so forth. If someone else does the guards, be sure to point this section out to him.

Homemade guards are rarely an aesthetic delight. Usually, they are boxes of 16-gauge steel hinged along one edge and with a latch or fastening opposite. Sometimes there is no latch: gravity holds the guard in place. What would happen if you beat on that guard with a wrench? Well, in any machine, you can expect some vibration and where there is a heavy reciprocating action (flatbed presses, bag machines, stamping, and a host of others), the impact between the free edge of the guard and the surface of the machine is not very different from a banging wrench. It is not at all uncommon for guards of this sort to be the worst noise sources on a machine.

To start with, you can use a heavy mesh of expanded metal instead of cold rolled steel. Even if a guard of this sort is rigidly fastened to a surface that is vibrating intensely, it doesn't radiate much noise. The individual strands are so small that they can't make waves long enough to be in the audible range.

Where sheet material is required (and it may even be required as a noise

barrier), a gasket, felt, or tough vinyl or neoprene foam between the mating surfaces will kill much of the impact noise. If you need such material quickly,
you'll find the foam weatherstripping tape sold at the local hardware store will
do in a pinch. More durable materials are available.

Don't be fooled when you select a foam for this use. The ease with which you
can compress it between your thumb and finger is deceptive. You may think
that it will be just as easy to compress it from $1/4$ to $1/16$ in. on three sides of a 2 ft^2
guard. There are many running inches there—you may find you have to sit on
the guard to get it closed! Two useful ideas in setting the clearance you will
allow for the gasket: see what deflection is produced by the loading that will actually be imposed (you can often do it best using a postal scale and a
straightedge to apply the force); check the manufacturer's literature to see what
it says about "permanent set."

PROBLEMS

1   A 1 hp 1750 rpm motor mounted on a concrete pedestal carries a 4 in.
    diameter V-belt pulley. The belt is driving a centrifugal pump with a 2 in.
    pulley. The fluid is clean water at room temperature. Vibration at the motor
    has caused one of the mounting bolts to crack. What sort of vibration con-
    trol is indicated? Think before you read on!

Answer

None of the methods described in this chapter! There is no reason that vibration
this intense should develop in an electric motor driving a steady load through a
belt. The motor is out of balance or has already chewed up its bearings.

2   A welding generator is powered by a four-cylinder, two-cycle engine that
    typically turns 2800 rpm. Two candidate vibration isolators for mounting
    the engine-generator are under consideration. One is an excellent patent
    mount that will achieve a natural frequency of about 3 Hz. The other is an
    inexpensive rubber pad that will produce a natural frequency of about 18
    Hz. Is the more expensive isolator likely to be worth the cost premium?

Answer

Two-cycle engines produce a power stroke from each piston in each revolution.
The lowest exciting frequency will, therefore, be

(2800 rpm $\times$ 4 strokes per revolution)/60 sec per min $=$ 187 Hz

The ratio of driving frequency to natural frequency for the cheap pad is 10.
Unless there are some other very special considerations, the ratio produced by

the better isolator (60+) cannot offer much better performance than the pad can.

3 A stacked screen sifter operates with seven impacts per second. Which of the isolators in problem 2 can be used?

Answer

There is no way to tell from the data you have. The natural frequency is determined in part by the loading. Unless the isolators under the sifter are loaded exactly as they were under the generator, there is no way of knowing the natural frequency.

12

DAMPING

Experienced people in noise control work are careful to distinguish between *damping* and *dampening*. It's true that they might dampen the floor before sweeping up to keep the dust down. When they set out to quiet a ringing part, they always *damp* it—sometimes by using a *damping* material and sometimes by some mechanical arrangement that *damps* the system.

Most people have a poor understanding of what damping is. By now you probably have a good idea of what it means. We are not going to pursue the theory, which recognizes several kinds of damping and produces a lot of heavy math. We will stop to bow to two of the ways a vibrating system is sometimes damped.

Try two mental experiments. In the first, we'll hang a long coiled steel spring from the ceiling over there and fasten a 2 lb weight on the bottom. Now mentally, pull the weight down a few inches and let go. What happens? It is bobbing up and down and still traveling through about the same distance on each bob. Now stop it for a second, unhook the weight, and slip a long, light paper tube over the spring. Hook that weight on again and give another pull. Aha! Look, it's slowing up already.

The second experiment requires a little more imagination. Our absent-minded lab technician left the stirrer in the rubber latex tub last night and the latex coagulated to a nice, bouncy, elastic rubber. The stirrer motor was off, so that's okay. When you try to get the stirrer blade out of the rubber, what happens? Every time you pull it to one side and let go, it flies back and forth . . . boing-oing-oing, right? Suppose the blade were in lard, or cup grease, and you pulled it to one side? Squish, once, right? And no boing.

Both of these experiments are dramatic examples of damping. And they both illustrate a key point about trying to control noise with damping. These were both naturally resonant systems. Damping worked to control the vibration in them. Now if the weight had been bobbing up and down because the ceiling was moving up and down 3 in., or the stirrer blade was moving because the tub was rocking back and forth, the paper tube would not have had any effect and it wouldn't have mattered whether it was rubber or lard in the tub. Damping makes no impression on a system in forced vibration. (It can help if the system wants to resonate at a harmonic of the forcing frequency.)

In both cases, energy was robbed from the damped system by something that, in the final analysis, must be called friction. The turns of the spring sliding over the paper tube are obviously sliding friction. The deformation of the lard is friction of a kind, too. Obviously, if you deformed it vigorously enough and long enough, it would get hot. Thus the energy was finally going into some kind of friction mechanism.

WHAT'S DAMPING GOOD FOR?

"The Cadillac door sound" was a piece of advertising jargon a few years back. Slam the door on a Cadillac and you hear a solid, quiet thud. (Supposedly, if you slammed the door on a brand X car, you'd hear a deafening, tinny bash.) Actually brands X, Y, and Z (and even F, C, and P) were using the same trick that Cadillac was using. The doors didn't sound good because they were good. They sounded good because they had a 4 × 4 in. chunk of asphalt-impregnated felt stuck to the center of the inner surface of the metal door panel. There is damping—via the lard route—in action. You'll find it, if you look, in steel office furniture, your typewriter, under the hood of your car, and the fact that the dealer always suggests undercoating (to make the car sound quieter).

In industrial noise control, there really isn't that much application for damping. That is not to say there is never a time and place for it.

Case 1

A coffee manufacturer put up a new plant several years ago. It was laid out to make use of gravity flow. The beans flowed to the scales in metal chutes. After roasting and cooling, they dropped another floor to the grinding operation in chutes. And then to the next floor. And on and on.

This was long before OSHA noise regulations. Even without legal requirements, though, this plant knew it had a problem—employees were going bananas with the hiss of beans in chutes. It happens that they used a noise control method which incorporated damping along with other techniques. They might have solved the problem at lower cost with damping alone. (Note: The bean impacts came at random. The worst of the noise was the resonance of the steel faces of the chutes.)

Case 2

In making paper for magazines, one step is to coat the sheet with a slurry of clay to make it white and opaque. This is applied in various ways but you can imagine that not all of the coating gets on the paper.

In order to keep gobs of it from traveling with the roll driving the paper, a doctor blade (about 0.04 × 2 in. × 20 ft. long) is clamped to scrape the roll. The roll has a surface speed of about 5000 ft/min and the blade is superbly hardened steel. Can you see where damping might fit in? Is there a resonant system lurking here? What vibrates and what drives, and what can be done about it?

That hard steel blade resonates up in the 8000 Hz octave. The roll drives it. Two ways to damp the blade were suggested. One was to tack weld a used blade to it. Spot connections between two sheets make a much more highly damped system than a single homogeneous sheet. The other was to mount the blade in a "lossy" rubber boot. (One of the favorite materials in the "lossy" category is butyl rubber.)

Case 3

Think of the discussion of noise from sheet steel guards that ended Chapter 11. Now it is perfectly true that the noise made by the guard as it hits the frame has the same pulse or repetition rate as the noise of the machine itself. The spectrum produced by the guard differs from that of the machine. The way the power is distributed with frequency is governed by the geometry of the guard. In other words, the guard rings.

With rare exceptions—and you would spot them quickly—guards make terrible bells. They are not very efficient resonators. The set of tones they produce is not musically consonant. None of that matters nearly as much as the fact that they do resonate. That fact gives you a chance to control them by using damping.

Figure 85 is a representation of the sound waves produced by a bare steel guard and the same guard after damping has been added. It is obviously a

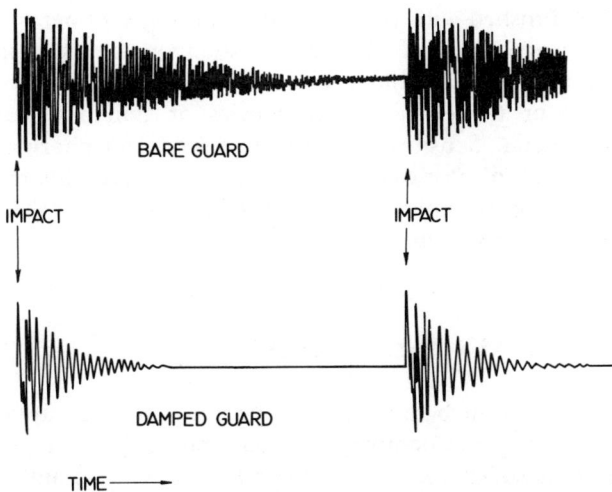

Figure 85   When a steel guard or any other ringing object is damped, the noise radiated because of its vibration after impact decreases. It is typical that the higher frequency vibration is the first to decay.

drawing and not a photo of an oscilloscope display but it shows general features you would expect to see if you fed the output of a good sound level meter to a scope.

1 Each impact produces a train of waves that slowly decays. The rate of decay for the damped guard is higher because the damping is draining the vibratory energy out of the system.

2 The bare guard shows quite an irregular waveform rich in high frequencies. This is typical of a part ringing at several frequencies that are not harmonically related.

3 The damped guard generally shows lower frequencies but it starts with the same sort of mixed-up clutter. This quickly drops out and only the lowest frequency remains.

Typically, the change in decay rate is greatest for higher frequencies when damping is used. The faster decay insures that the average level is lower. The greater effectiveness of damping with high frequency will change the "clash" character of the noise in the direction of becoming "thunk." While you may not be interested in the Cadillac door sound in your industrial equipment, you should be interested in the relative importance of these frequencies after A-weighting.

The most common method of damping such a guard is to line it with a composite material that will place a thin ($3/16$ to $1/4$ in., usually) layer of lossy foam or glass fiber against the surface. The next layer of the damping composite is usually either 1 psf lead or a comparably heavy layer of limp sheet of barytes in a mastic or rubbery binder. If the damping is being applied to the inside of the guard, it is often finished with film covered foam or glass fiber to absorb noise.

Three other methods often used in this situation are the application of a limp layer of self-adhesive lead-loaded vinyl (-325 mesh lead powder in well-plasticized vinyl) or trowel-on loaded mastics or building the guard from laminated sheet metal. Sensitive products like food and pharmaceuticals preclude the use of lead in most cases. There are some acceptable trowelable mastics for such use. For example, you can make your own by mixing clean sand and PVA adhesive ("white glue").

DAMPING MATERIALS AND METHODS

Though a long way from being the most pressing problem area in industrial noise control, damping and damping materials have a fascination.

The cheapest material you can consider for damping is automobile undercoating. It's not very effective, but it's cheap. Slather away, damping fans. Damping becomes much more effective if you can find a way to load it up with something massive. A ringing gear, for example, wouldn't yield an inch to a heavy coating of automobile undercoat on its cast iron spokes. However, if you wrap the coated spokes with a layer of ⅛ in. lead (8 psf, it's available at your local plumbing supply house), and clamp these in place with radiator hose clamps, it may just stop ringing.

There are a host of proprietary damping compounds. However, the respective merits of these compounds is less important in damping than whether you are lucky enough to put the right amount of the right stuff in the right place. All the compounds work in *some* situations, but it is a trick to find a cheap application that is foolproof. If you like to tinker, you'll love noise control by damping. The materials themselves range from troweling compounds to sheet materials that are self-adhesive to sheet materials that are supposed to be held in place by riveting through retaining straps or sheets, to casting compounds. In the latter case, you build a form around the offending object and pour in the damping material like concrete.

A caution in applying damping materials is that their damping properties *vary dramatically* with temperature. What you think is a good actor at room temperature with the machine shut down probably will not work the same way at 200°F with the equipment operating.

You can also purchase, for new construction, laminated metals. These have a layer of damping material between two sheets of metal. They come at a premium cost, but they can save you money. Think of these when you have many units to build and can take the time to wring out a prototype first. In heavy construction, you'll have a problem with these materials—they cannot be welded easily. The damping material is destroyed by high temperature.

SAW BLADES

Of all the successes in using damping for noise control, none is more common—or more striking—than the use of damping sheets to quiet circular saws. Analyze the sources of noise at play when you run a sheet of plywood through the saw:

1 Individual teeth of the blade are impacting the sheet at a well defined rate.
2 The blade is tearing and shearing fibers—surely not without making some noise.
3 The teeth and gullets are slicing through the air at high speed and producing aerodynamic noise and perhaps even some edge tones.
4 Worse still, they plunge into the kerf in the wood and emerge again at explosive speed. This is a fairly effective noise generating mechanism.
5 The light sheet of plywood is big and stiff. Under regular impacts, it must be radiating a lot of noise.
6 In fact, it may even be bouncing on the table. After all, a great deal of power is being applied to it.

In the light of such an analysis, there can be little wonder that circular saws are so noisy. However, the analysis leaves out at least one key noise source and, believe it or not, that ringing saw blade is *the* major source of noise in most cases.

A mental experiment and a little analysis of the physics help here. For the experiment, stop the saw, remove the guard, and tap the blade with a wrench. Does it ring? And how! Look at the blade in your mind's eye now. It is a fine piece of steel—that's what you paid for. It is, therefore, highly elastic. Moreover, it is symmetrical about its single point of support on the arbor. It is even clamped very rigidly there by the arbor nut so that there are minimal energy losses when it rings. You could hardly design a better mechanical resonant system if you tried.

You need take nobody's word for the fact that at least half the noise comes from the ringing blade. Adhering a thin sheet of damping material to the side or sides of the blade almost always reduces the noise of sawing by 3 to 8 dBA. The most usual figure is 5 or 6. For bigger blades, even greater reductions have been shown.

CONSTRAINED LAYER DAMPING

Adhering a damping material to a surface is the usual treatment. The laminated metal sheets mentioned previously and a treatment of sawblades by sandwiching the damping material between the blade and a metal plate are examples of constrained layer damping. Not only does this make efficient use of the damping material but it protects it in situations where it might otherwise be damaged.

These brute force situations are fairly common. In conveying logs, for example, it is usual to change the direction of flow by letting them fall into a trough made of steel plate with a chain conveyor at the bottom of the trough. Rocks or heavy castings falling into a steel hopper is another example.

Where impact and abrasion are not this severe, you can consider a bonded rubber lining that will furnish some damping and also will absorb some or most of the energy of impact. This, of course, is exactly the problem. When half-ton logs hit the rubber after falling 8 ft, shredding and tearing of the rubber is certain. instead, use a layer of damping material on the back side of the steel plate with either steel straps or a second plate to hold it in contact.

A typical construction starting with a ½ in. steel plate would be a backing of ¼ in. of tough damping sheet and ¼ in. of steel plate behind it. The lamination may be cemented together both as an aid in assembly and to assure contact at the interfaces. Do not depend on the cement as the sole fastening! A much better practice is to rivet this construction using countersunk flush heads on the ½ in. steel side.

As in all other exercises in building noise control hardware, you simply must consider the possible consequences of your design. The method just outlined would be fine for a log conveyor—*as long as it does not lead to the chipper!* A broken off rivet head dropped into that high speed knife will become a piece of lethal shrapnel. Go ahead with the scheme if there is a tramp metal detector downstream. In other cases, however, you might find no way to keep the operation safe. What would happen if the rivet head got into the feed of a screw extruder?

There are alternative designs, such as welding studs to the back of the original steel. Your own ingenuity or advice from some of the seasoned veteran engineers in your plant can find a safe way of implementing damping designs in these brute force situations. This is not the point of the message here. What you should remember is that the same punishing forces that made you put the damping on the safe side of the steel will be trying to demolish the hardware you install. And they may succeed. Plan your installation so that nothing terrible will happen if they do.

NONSTRUCTURAL APPLICATIONS

Damping, as a noise control measure, is usually meant to imply the sorts of mechanical systems described here. In a broader meaning, it applies to tuned electrical and electronic circuits and, in fact, to any resonant system.

Aithough industrial noise problems involving organ pipe phenomena are not everyday occurrences, when they do occur they are often outrageously noisy. One example, given in Chapter 1, involved an industrial vacuum cleaner with a duct that just tuned the vane passage frequency of the blower. The easiest solution to that problem was to change the duct length. However, another solution worth considering would have been to line one side (or any part) of the duct with absorber.

This is not a trivial "bandaid" noise control; it is fundamental. You should not be thinking in terms of an absorber to reduce sound power in the reverberant field here but as something that is draining energy out of a resonant system. Ah! The absorber damps the oscillator, possibly to the extent that the system can no longer resonate.

For a final example, look at Figure 31 again. What is the glass fiber doing in this system?

ESTIMATING PERFORMANCE

Calculation methods for estimating the effect of damping exist. They are complex—literally and figuratively. Even the modulus of the elasticity must be treated as a complex number (*i* and all that). You will be taking on a lot if you attempt to make such calculations. The first step is the worst—getting your hands on data that mean anything. Unfortunately, the standard procedure with damping is "try it and see"!

SOMETIMES DON'T DAMP

The key question, when damping is up for consideration, is whether there is a resonant system. If nothing is resonating, damping won't help. Think of a familiar analogy of resonant systems. Suppose you were annoyed with the opera (rock music, whatever) that was coming over your radio. Would your first thought be to stuff wet towels into the loudspeaker so that you wouldn't hear it so plainly? Hardly. You'd tune to another station. (You might even turn if off, but that doesn't serve the purpose here very well.) What did you do when you tuned another station? You shifted the frequency at which the tuning circuit of your radio resonated. Did the station you didn't like stop broadcasting? But you didn't hear it either, did you?

This technique is worth considering in industrial noise control, too. It is the first thing that ought to occur to you if a panel, or a guard, or anything wants to resonate at some frequency being produced by a set of gears or any other source. Tune the panel to a new frequency. The two quantities that do this best are mass and stiffness.

For more (or less) mass, you can replace the panel. You can trowel on, or rivet on, anything, or sometimes you will have to take heroic measures and cast concrete against it. For stiffness, you will weld on angles or channels or screw on some stiffening battens of wood. Take care (and abandon the tidy habits that got you into engineering) that the stiffening goes in at crazy angles and spacings. If you build a new panel with perfect symmetry, you may find it still likes to be excited by the same source—this time at a higher harmonic.

13

BRINGING THINGS TOGETHER

PLANNING A NOISE CONTROL PROGRAM

Let's start by thinking of OSHA as an advocate and not an enemy. The enemy is surely noise. In the same way that the Internal Revenue Service expects to be questioned and advises you to make use of every legal method of reducing your taxes, OSHA expects you to question its findings, offer supported evidence that it may be in error in some citation, and to obey the law as it is written and interpreted. OSHA does not expect you to accomplish some vague "good thing" in controlling noise.

Another object of noise control may occur to you: the conservation of hearing, either as a "good thing" to accomplish, or more selfishly, to prevent a lot of compensation cases piling up for your company in the future. Let's put that aspect off for a while because, as will become apparent, meeting OSHA's requirements is the more tidy goal. Hearing conservation, by means of a noise control program, is not yet a very well defined exercise.

WORKING WITH OSHA

A True Story

The manager of a noisy plant was explaining the situation to his noise consultant. "The OSHA inspector introduced himself, showed me his credentials, and then told me that the penalty for hitting an OSHA inspector is 10 years in a Federal prison," the manager said.

You may well hope that when the OSHA inspector comes to your plant the conversation doesn't take that turn. Certainly *you* don't want to make him have those thoughts. The very first thing you should do in working with OSHA is to show the inspector that you want to deal with him professionally.

You can be courteous. However, this is not nearly so important as being professional, nor does any of this mean that you can't disagree with him. When you do, disagree about the substance of what he is saying. And do it professionally. Don't ever convey the impression that you think the requirements of the law are nonsense . . . or that your management thinks they are. You are going to make your life very difficult if you do. If you or your management have these strong convictions, convey them to your Senator or Congressman.

There is a real point in working *with* OSHA. They know you may be in a very difficult position in trying to control noise. Your reasoned approach encourages them to be reasonable in understanding that difficult position. If, on the other hand, you are emotional and make a personal fight of it, what sort of response will that induce? There is a good chance you will end up proving (?) your case to a very impartial arbiter indeed—a Federal District Court Judge.

If your job is to meet OSHA's requirements, do it effectively and at least cost. Don't make your job harder in the first inning. And don't let your management put you in that hole either.

RULES OF THE GAME

There is quite a common notion of what OSHA requires in terms of noise control. Many people think that nothing in the plant can be noiser than 90 dBA to satisfy OSHA. It's just not true! Under the 1971 law OSHA has set an exposure time that is acceptable for any level between 90 and 115 dBA as long as it is the only exposure to noise at a level above 89 dBA that an employee has in his working day.

Table 11 shows what the law demands. For an 8-hr working day it is true that a continuous 90 dBA uses up the whole noise exposure permitted. But if the equipment sometimes stops, or some of it, it is worth noting that 91 dBA is acceptable for 6.96 hrs, or 92 for 6.06. In fact, 115 dBA is acceptable for 15 min as long as there is no other exposure above 89 dBA.

The nomograph of Figure 86 will give you the fraction of the permitted exposure that results from any time and level within the limits. (It can also be used with a proposed 1974 OSHA standard that accounts for levels down to 85 dBA and exposure times as long as 16 hrs a day. Although it's unlikely that many people regularly work 16 hr days, working days longer than 8 hrs are common. Many companies have elected to base their current noise control work on the 1974 criterion so that any future tightening of the law will not undo their efforts.)

The fact that the 1974 proposed law takes levels between 85 and 90 dBA into account is more important than you may at first realize. In a sizable portion of most plants, these are the usual noise levels.

Most, if not all, industrial noise situations involve exposures to more than one level of noise. In this case you must tally up the total exposure that results from all the individual exposures for each employee in an average day. Let's call C the duration, in hours, of exposure for any employee at a location whose level

Table 11 *Permitted Exposure Time*

Level (dBA)	Hours of Exposure Permitted	
	Existing (1971) Law	Proposed (1974) Law
84	No limit	no limit
85	No limit	16.00
86	No limit	13.92
87	No limit	12.12
88	No limit	10.56
89	No limit	9.18
90	8.00	Same as 1971 law
91	6.96	Same as 1971 law
92	6.06	Same as 1971 law
93	5.28	Same as 1971 law
94	4.59	Same as 1971 law
95	4.00	Same as 1971 law
96	3.48	Same as 1971 law
97	3.03	Same as 1971 law
98	2.64	Same as 1971 law
99	2.30	Same as 1971 law
100	2.00	Same as 1971 law
101	1.74	Same as 1971 law
102	1.52	Same as 1971 law
103	1.32	Same as 1971 law
104	1.15	Same as 1971 law
105	1.00	Same as 1971 law
106	0.87	Same as 1971 law
107	0.76	Same as 1971 law
108	0.66	Same as 1971 law
109	0.57	Same as 1971 law
110	0.50	Same as 1971 law
111	0.44	Same as 1971 law
112	0.38	Same as 1971 law
113	0.33	Same as 1971 law
114	0.29	Same as 1971 law
115	0.25	Same as 1971 law
116	None	None

The exposure time permitted supposes that there is no other exposure adding to the noise burden.

16
12
10
8 *
6
5
4
3
2

DURATION OF EXPOSURE

1
60
50
40
30
20
15

MINUTES

•EXTENSIONS WOULD BE
ADDED UNDER 1974
PROPOSED REGULATION

85
* 90
95
100
105
110
115

NOISE LEVEL dBA

NO EXPOSURE
AT HIGHER LEVELS
IS PERMITTED

.05
.06
.08
.1
.15
.2
.3
.4
.5
.7
1.0
1.5
2
3

FRACTION OF PERMITTED EXPOSURE

OVEREXPOSURE

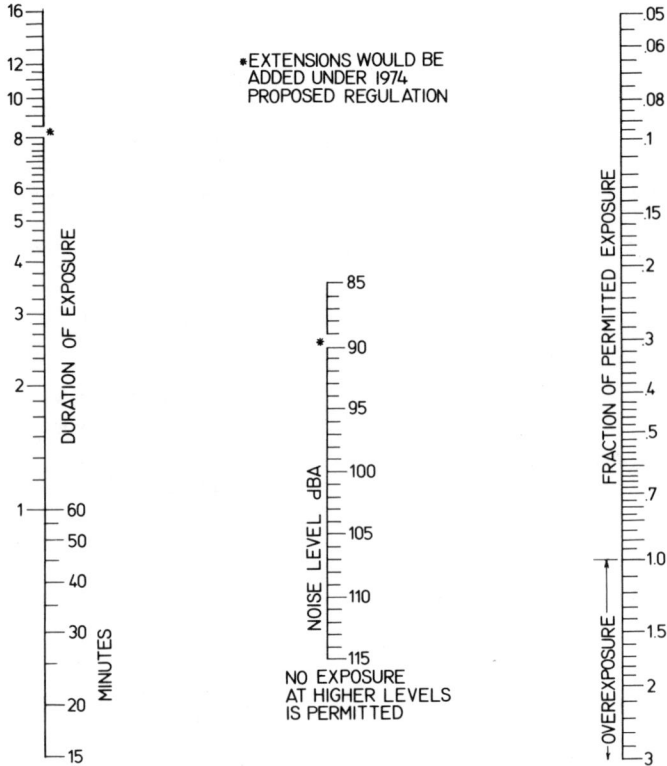

Figure 86   Permitted exposure nomograph.

can be well defined. We will call T the permitted exposure time, in hours, for the level. What OSHA wants you to accomplish is to have

$$C_1/T_1 + C_2/T_2 + C_3/T_3 \ldots C_n/T_n = \text{no more than } 1.0$$

As we shall see, this is very different from "bringing everything in the plant down to 90 dBA or less."

In the first place, in some worked examples that follow, you'll see that it is quite common to be able to leave locations at which people work at levels well in excess of 90 dBA. This usually means you save money.

In the second place, cost sometimes can become huge in attempting to reach 90 dBA. In fact, for some operations, there may be no way to get the noise to 90 dBA and still retain a practical process. These difficult operations are not exotic things that happen once in a while in some strange manufacturing process. Do you have big boiler feedwater pumps, lobed blowers, fans with high static pressure requirements, circular saws, cleaning by steam or high pressure air? There are so many more of these it is not worth going on. The point is that one or more of these noise sources exist in your plant or plants.

In an age when we can land men on the moon and bring them back, we can no doubt get these sources to 90 dBA. Often, however, it will require remote control—with TV monitor cameras, perhaps—or automation of complicated and nonrepetitive jobs (one-of-a-kind crates to be gotten out on an automated saw, for example). Finally, what about maintenance adjustments that must be made while noisy equipment is operating?

Can you imagine an AR-2-DEE-2 maintenance robot in your lifetime? (Nor should you suppose that OSHA would be delighted with that idea, since OSHA is part of the Department of Labor.)

The difference between "everything below 90 dBA" and the sum of all the C/T terms being less than or equal to 1.0 ought to be clear. Now you have two simple, straightforward questions to answer.

First, when should you undertake noise control? You can wait until OSHA cites you, or you can go exploring for yourself and begin to get noise under control when you find that it is causing some employee to have a total C/T of more than 1.0. You really shouldn't wait for the citation. OSHA inteprets this as "a willful violation of the law."

Second, how much noise reduction should you bring to bear? What noise level should you shoot for as a target? That's easy, too: a noise level that results in a *total C/T* of 1.0 or less for all employees.

No, that's not so easy! The noise control part may or may not be difficult, but finding the best levels for all locations that cause excessive noise exposure (in the legal sense) for all employees is a complicated problem.

SURVEYS AND GRID SURVEYS

Historically, a way of coping with that second question was to go exploring in the plant until you could mark a noise level on the floor plan for every location in the plant. This was called a survey.

The difference between a survey and a grid survey depended on how seriously you took the word "every." In order to be thorough, the grid survey uses a tape measure or snapped chalk lines on the floor to pinpoint the locations at which sound level readings are to be taken. They are taken on 10 ft centers or sometimes less closely.

The idea is to be able to draw reliable noise contours. This frequently means returning to the measurement points so that cases involving steep gradients can be resolved.

This logical beginning leads to a logical question. When you have completed the survey (grid survey, same thing), what do you know? Obviously, you know many sound pressure levels. And then . . .? The point is that you are not even close to knowing the total C/T of each of the workers in the plant.

DOSIMETERS

A recent—and in some people's view, perverse—answer to this has arisen with the notion that the first thing to do is to have your employees carry an integrating sound meter called a dosimeter. You check this each day for a week (more for people in the maintenance department, since their duties vary considerably from day to day). From the data you collect this way, you can tell, eventually, which peope have C/T's of more than 1.0. Great! Where did they get it? Although it may be easier to answer this question than it was to interpret what you had when you had completed a grid survey, you still don't have the data you need.

This isn't the only trouble with dosimeters, incidentally. You can skip the clowning or even the malicious employee who takes advantage of the novelty of wearing the fancy gadget the front office wants to try out. If he wants you to get funny answers, you surely will.

The instruments themselves have not had a good record. The University of Pittsburgh tested several on a rotating rack in a reverberation room under carefully controlled (laboratory) conditions. The manufacturers of the dosimeters had all been fully informed of what the test would be and were asked to furnish instruments in good calibration and suitable for such testing. The results ranged from about 70 to 150% of what the actual noise exposure had been.

Moreover, *that* isn't the only trouble with dosimeters. What would you do if the daily C/T readings for one employee, doing the same task, were 0.65, 1.9, 0.95, 0.72, 1.65 at the end of a week? Was he doing the same thing each day? What will you take as his average exposure? Do you want another week of data? Do you want about 10 times as many dosimeters so that you can work your way through the whole plant population? Do you really want 10 times as many dosimeters?

THE OTHER STARTING PLACE

Now let's say we have every faith in the dosimeter readings. When we have found an employee with a consistent C/T of more than 1.0, what comes next? One thing that surely occurs to you is to find out where he spent his day. If you can learn this, and have confidence in what you have learned, you can begin to make use of the survey data.

Now this is supposed to be a book about industrial noise control and the homilies and moral object lessons will be held to a minimum. But if you think that grid surveys are prone to an occasional error and that you might have doubts about dosimeter readings, look out!

When you try to find out where an employee spent his day, he will tell you that he spent the full 8 hours at his machine, running it flat out. His foreman will tell you that they average less than 5% downtime. The production superintendent will also defend the efforts of his department. Nobody ever goes to the men's room, you will have to remind people about lunch and coffee breaks, ask if they ever had to get a replacement part or call for a maintenance man, and so on.

A great idea is to ask what the normal productivity of the machine is in units, parts, pounds or feet per hour and then check into the actual production for a typical month.

A survey alone, or dosimeters alone, or both in combination lead you back to the same central question of where people spend their time. The other starting place in a noise control program is to find out where they spend their time *before you do anything else.* This is the same difficult job just described. Cross check in every way you can find. Second line management is probably the best place to start. The department head should have a good idea of what his people are doing and the good sense to ask a foreman when he is in doubt.

If you are tactful, you may be able to get him to check the production records against the nominal hourly production rate of the machines—and sometimes he'll even thank you for what he has learned in the process.

When you have a list of where each employee spends time, and how much time, you will know all the places where you must measure noise levels. The first bonus in this procedure is that you will not measure nearly as many locations as you would have in a grid survey—nor will you have to wonder how to draw contour lines.

STRATEGY AND TACTICS*

Because all of us engaged in the area of industrial noise control tend to be fascinated with ways of winning the practical battle of quieting a specific piece of equipment, we tend to lose sight of the more important goal: the strategy of winning the war by meeting requirements for the whole plant at minimum cost. What follows may be less fun, but it does reveal a way to win that war at minimum cost. In several cases it has been possible to compare this technique with plans to bring the noise level to 90 dBA throughout the plant. This technique has always saved 40%, or more, of the cost of the 90 dBA approach.

The savings come from, first, making noise reductions at as few places as possible to achieve a C/T of 1.0 or less for all workers, and second, from setting the minimum noise reduction at each of these locations. There is a perfectly routine method for doing this.

There is some justice to an objection that achieving even a few decibels of noise reduction at some locations will be very expensive while quite substantial reductions at other locations may be available economically. To trim 5 dB from the noise from a big reciprocating compressor is going to cost thousands of

*This and some following sections are largely taken from an article by Thomas D. Miller and the author that appeared in *Sound & Vibration* magazine; September 1977. Portions from that article are adapted by permission.

dollars, at least. Yet, screwing in a $5 muffler may scalp 25 dB from the noise of an air jet out in the plant. For the moment, though, suppress your hunter-killer tactician instincts and take the cool, reasoned view of the strategist.

Consider an extremely simple case: An employee is exposed to noise at only two locations in his typical working day. He spends 4 hrs near that reciprocating compressor at a level of 95 dBA. He also spends 15 min a day at another location where an air jet produces 115 dBA. Either location uses up his entire permitted exposure for the day. Note that *either* location could be controlled to achieve C/T of 1.0. Note that *both* locations could be reduced by lesser amounts to achieve the same thing. Nobody will blame you for choosing the $5 solution in this simple case. But then, it is the *simple* case.

Consider a complication: 10 other people work near the compressor and they have C/T figures ranging from 1.2 to 1.4. Ah! There is room for strategy, you see.

TAKE THE LOFTY VIEW

Forget the plant hardware now and look at your problem from the loftier view of a noise control strategist. What is it that you have to do first?

1 Find the C/T for each employee in your plant.
2 Separate all those whose total C/T exceeds 1.0.
3 For those who have C/T of more than 1.0, find the contribution of every work location they occupy in their working day.
4 Regroup your data and find out which work location accounts for the greatest C/T to all these employees.

This is a succinct statement and it may bear rereading or marking if you have the habit of marking the meaty sections of your books.

Let's do that again—in slow motion—for something really out there in the plant. It is still a very simple case, but illustrative of how to apply the four directions suggested. Suppose just three employees are operating a semi-automatic unit producing something or other. The equipment is shown in Figure 87. The three employees, an operator, his or her assistant, and a helper, are shown in the locations they typically occupy. Annual production records show that this machine operates 55% of the time.

You have interviewed the department foreman and learned that the operator is away from the machine for half an hour a day in the men's room or on coffee breaks. You've also learned that he spends 1.7 hr at the control panel. In fact you've persisted in interviewing the foreman, and checking production records, and so forth, until you can tabulate exposure times, locations, and associated levels (Table 12).

In the same way, with quiet, persistent, and always tactful questioning, you

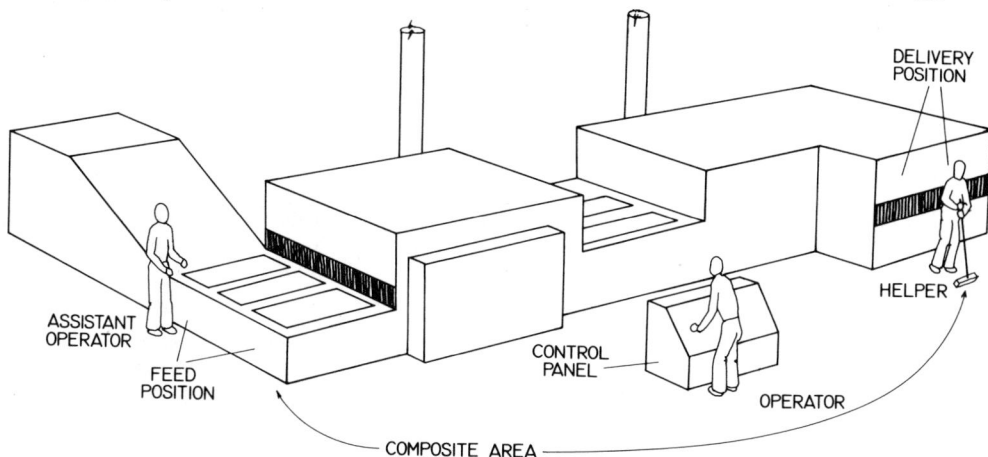

Figure 87 Four exposure locations around this hypothetical machine serve to define the exposure of the three employees who operate it.

Table 12 *Operator's Exposure Profile*

Locations	dBA	Max Time Permit	Duration	*C/T*
Control panel	99	2.30	1.70	0.74
Feed position	102	1.52	1.00	0.66
Break area	81	∞	0.50	—
Machine off C.A.	91	6.96	3.15	0.45
Machine on C.A.	95	4.00	1.45	0.36
Steam cleaning	109	0.57	0.20	0.35
Daily noise dose				2.56

have learned that the assistant and helper spend their day as shown in Tables 13 and 14.

In the interview, for example, you have pointed out that the machine only operates 55% of the time. "What happens when it is down?" you asked. This question produced some important information—the machine is regularly, and noisily, steam cleaned. You may also find that the operator spends a lot of time talking to his friend in the tool crib. Tact, now, is very important: but so is *fact*. If you find a fact like this, include it in your exposure analysis. A real representation of long-term exposure is essential and the real facts may be far from what a casual inspection would indicate.

In the tables you have developed, the abbreviation "C.A." stands for "composite area." It is typical of a class of exposure locations that can be found in any plant. Inspection tours and trips to the tool crib or front office are the common ones. (In order to assign a noise level to them, you will need to use a

Table 13   *Assistant Operator's Exposure Profile*

Locations	dBA	Max Time Permit	Duration	C/T
Control panel	99	2.30	0.38	0.17
Feed position	102	1.52	1.00	0.66
Delivery position	107	0.76	1.07	1.41
Break area	81	∞	0.50	—
Machine off C.A.	91	6.96	3.15	0.45
Machine on C.A.	95	4.00	0.70	0.43
Steam cleaning C.A.	109	0.57	0.20	0.35
Daily noise dose				3.47

Table 14   *Helper's Exposure Profile*

Locations	dBA	Max Time Permit	Duration	C/T
Feed position	102	1.52	0.50	0.33
Delivery position	107	0.76	1.45	1.91
Break area	81	∞	0.50	—
Machine off C.A.	91	6.96	3.15	0.45
Machine on C.A.	95	4.00	2.20	0.55
Steam cleaning Oper.	112	0.38	0.20	0.53
Daily noise dose				3.77

statistical method. There are instruments that will do this for you. You can also do it with a sound level meter—see the first worked problem in Chapter 5, the 1-2-3 method at work.)

STRATEGIC PRIORITIES

Thus far you have worked your way through the first three suggestions. Now you need to regroup the data and find which location should have top priority for noise control work. To do this, find what *fraction of the total C/T* for each employee is contributed by each location. For example, the feed position contributes to all three employees' *C/T*. For the operator it contributes 0.66 of the total 2.56 he acquires in a day. In the same way, the assistant gets 0.66 of his 3.47 from it and the helper, 0.33 of his 3.77. Summing all these, the exposure index for the feed position is

$$\frac{0.66}{2.56} + \frac{0.66}{3.47} + \frac{0.33}{3.77} = 0.54$$

This number 0.54 has no absolute meaning. But it does give us a handle on the extent of the problem caused by the feed position. When you have computed an index for each position, rank them according to importance to produce your strategic priority list (See Table 15).

The original level is included in Table 15 because, having established the priorities, the next step is to set the noise reduction required at each location. That reduction is usually called the *design goal*. Look over your priority list, now, and mark off any location where a $5 muffler will do the trick. Also mark off any location where no reasonable amount of noise control can be brought to bear. In the first case, set the design goal as 25 dB (or whatever you can expect from the muffler) and in the second case 0 dB. None of the locations in this list have so easy, or impossible, an outlook.

DESIGN GOALS

In this list, all things are possible and no things are preferred. It is a model case. The next exercise, therefore, persists in the strategic approach—it imagines that the cost of noise control at each location is dependent only on the design goal. *If you know better, you will override the routine.*

There are many, many ways in which design goals could be found. The method used by the inventors of the technique is proprietary and involves using a computer with capacity for an abundant amount of data and state-of-the-art speed. Do not be discouraged. If you must do this work, limit yourself to departments, or even operations, that do not exceed about 50 employees. Break out all your operations so that they each have little contact with one another (in the sense that there will be no, or few, people who have exposure in more than one area). You will still want to keep all areas in some sort of priority order in dealing with them. However, for the first time you can begin to logically set aside an area that is slated to be discontinued . . . or to have a lot of new equipment added.

The next step is to make trial reductions, on paper, of the noise level and see

Table 15 *Priority List*

Location	Effect Index	Original Level dBA
Delivery position	0.91	107
Feed position	0.54	102
Machine off C.A.	0.42	91
Machine on C.A.	0.41	95
Control panel	0.34	99
Steam clean C.A.	0.24	109
Steam cleaning Oper.	0.14	112

how the fractional and total C/T figure for each employee responds. The method of doing this in a computer produces a sort of optimum result by reducing noise at the fewest locations and by the least amount. It is too tedious to consider in a calculation "by hand." Instead, an arbitrary procedure is offered. It has the virtue of speeed, but may ask for noise reduction at too many locations. A remedy for this shortcoming will be given later.

Use Table 16 to try noise reductions according to the level found at each location. Don't be slavish in following this table. Two common exceptions will occur to you immediately. First, don't call for noise reductions that you cannot get! It will not be productive to set a design goal that will achieve a level of 89 dBA at a location or near a local source when the whole area is at a level near 100 dBA (and, of course, the exception to that exception is that it will if you have in mind putting the operator in a noise control booth). Second, take advantage of the fact that big noise reductions *will* occur at locations having a low noise levels when there is only one source of noise and you are shooting for a big reduction by controlling it. In this case, try a noise reduction smaller than the table indicated for the higher level but take credit for that same reduction—if the situation warrants it—at all the locations that are controlled by that source.

Using design goals suggested by the table, your first trial reduction will result in the figures shown in Table 17.

With these trial reductions, list new figures for C/T for the employees. The outcome of this is shown in Table 18. In round numbers this table shows that you have cut the C/T in half, but you haven't reached the exposures that OSHA requires. A second trial reduction, at least, will be needed. Before you resort to the trial reduction table again, scrutinize the partial C/T figures. Sometimes one of them will stand out and you will see a way of solving your problem. At least keep looking at the possibilities. Again, don't be a slave to the trial reduction table.

If no exceptional possibilities occur to you, make a second trial based on the trial reduction table and you will have the results shown in Table 19 (p. 225).

Now these figures are very close to what OSHA wants. In many cases you may find that two of the three employees have had their C/T reduced to 1.0 or less by this stage. If that is the case, it may be a good idea to make more critical measurements of the noise levels, to check the exposure times at your locations more closely, or see whether some of the work locations you have assigned might be better defined by splitting them into more than one location.

Table 16   *Trial Noise Reductions*

Level (dBA)	Reduction (dBA)
90 to 94	1
95 to 99	2
100 to 104	4
105 to 115	8

Table 17 *Noise Levels After First Trial Reductions*

Location	Original Level (dBA)	Reduction	First Trial Level (dBA)
Delivery position	107	-8	99
Feed position	102	-4	98
Machine off C.A.	91	-1	90
Machine on C.A.	95	-2	93
Control panel	99	-2	97
Steam clean C.A.	109	-8	101
Steam cleaning Oper.	112	-8	104

It is pointless to discuss the morality of this arrangement. The more critical analysis should in no way be equated with "fudging" the data, and the advice that it is sometimes worthwhile to do a more critical analysis is in no way an invitation to cheat. Think of it this way: you could end up in court! The question is not whether you have taken a high moral stand, the question is whether you can support your data.

In the model case given here, suppose that there is no such easy way out. Then, by all means resort to the trial reduction table again. This time it produces the results shown in Table 20 and the design goals for compliance have been established.

A common outcome of such an analysis is that for some of your locations within the area, no noise reduction at all will be required. (See Table 21.)

One last thought on setting design goals: this has been a paper exercise (with the enlightenment that your experience brought to it). There is no reason in the world for you to follow the list of design goals slavishly *when you get into the engineering of the noise controls.* You may find that you made some bad assumptions or that there is an easy way of cracking what looked like a tough nut. All that design goals are supposed to do for you is to let you get on with the next step, which is . . .

SETTING UP A COMPLIANCE PROGRAM

Armed with your design goals, make another extended tour through the plant. The object, this time, is to envision what must be done to reduce the noise by the amount of your design goals. In fact, you will now develop a predictable itch to get started doing just that. That is a very good idea *in some cases.* Where the cause of the noise is worn or broken equipment or steam leaks (the perennial example), by all means suggest that the necessary work be done. It will be no waste of effort. That work is needed even without a noise problem.

However, unless you can point to *the* one or two noise sources and are sure

Table 18 *First Trial to Determine Design Goals*

Locations	dBA	Max Time Permit	Operator		Assistant Operator		Helper	
			Duration	C/T	Duration	C/T	Duration	C/T
Delivery position	99	2.30	—	—	1.07	0.47	1.45	0.63
Feed position	98	2.64	1.00	0.38	1.00	0.38	0.50	0.19
Machine off C.A.	90	8.00	3.15	0.39	3.15	0.39	3.15	0.39
Machine on C.A.	93	5.28	1.45	0.27	1.70	0.32	2.20	0.42
Control panel	97	3.03	1.70	0.56	0.38	0.13	—	—
Steam clean C.A.	101	1.74	0.20	0.11	0.20	0.11	—	—
Steam cleaning Oper.	104	1.15	—	—	—	—	0.20	0.17
Break area	81	∞	0.50	—	0.50	—	0.50	—
Daily noise doses				1.71		1.80		1.80

Table 19 *Second Trial To Determine Design Goals*

Locations	dBA	Max Time Permit	Operator Duration	C/T	Assistant Operator Duration	C/T	Helper Duration	C/T
Delivery position	97	3.03	—	—	1.07	0.35	1.45	0.48
Feed position	96	3.48	1.00	0.29	1.00	0.29	0.50	0.14
Machine off C.A.	89	∞	3.15	—	3.15	—	3.15	—
Machine on C.A.	92	6.06	1.45	0.24	1.70	0.28	2.20	0.36
Control panel	95	4.00	1.70	0.43	0.38	0.10	—	—
Steam clean C.A.	97	3.03	0.20	0.07	0.20	0.07	—	—
Steam cleaning Oper.	100	2.00	—	—	—	—	0.20	0.10
Break area	81	∞	0.50	—	0.50	—	0.50	—
Daily noise dose				1.03		1.09		1.08

225

Table 20 *Third Trial to Determine Design Goals*

Locations	dBA	Max Time Permit	Operator		Assistant Operator		Helper	
			Duration	C/T	Duration	C/T	Duration	C/T
Delivery position	96	3.48	—	—	1.07	0.31	1.45	0.42
Feed position	95	4.00	1.00	0.25	1.00	0.25	0.50	0.13
Machine off C.A.	89	∞	3.15	—	3.15	—	3.15	—
Machine on C.A.	91	6.96	1.45	0.21	1.70	0.24	2.20	0.32
Control panel	95	4.00	1.70	0.43	0.38	0.10	—	—
Steam clean C.A.	97	3.03	0.20	—	0.20	0.07	—	—
Steam cleaning Oper.	100	2.00	—	—	—	—	0.20	0.10
Break area	81	∞	0.50	—	0.50	—	0.50	—
Daily noise dose				0.96		0.97		0.97

Table 21 *Final Design Goals*

Location	Original Level (dBA)	New Level (dBA)	Design Goal
Delivery position	107	96	−11
Feed position	102	95	−7
Machine off C.A.	91	89	−2
Machine on C.A.	95	91	−4
Control panel	99	95	−4
Break area	81	81	0
Steam clean C.A.	109	97	−12
Steam cleaning Oper.	112	100	−12

that fixing them will bring the area into compliance, it is better to continue to analyze the situation.

The real point of your tour, this time, is to make an estimate of the costs and time required to bring the plant into compliance with the law. Your ability to do this well will depend partly on the skill you have developed in noise control. It will depend much more heavily on the size of the operation.

For example, one program of noise control has been in effect at one plant for years under a compliance program calling for 11 years of work and accepted by the Department of Labor (OSHA). The work is on schedule, but it has taken the efforts of several engineers and a big computer, initially, to correlate all the exposure data to generate a priority list and set design goals.

In fact, in the interim, plant expansion made a second evaluation necessary so the work is now based on a second generation of design goals in the original areas of the plant.

Your ability to compile a minimum cost program to achieve compliance will be limited by the staff and facilities available to you. You can do it quite easily for a small plant or a department. At somewhere around 50 people, you will probably need help.

Your final program should:

1 Set up a realistic schedule showing when all individual noise control projects begin and complete. The time periods allotted to engineering, management acceptance of the controls or selection of alternates, delivery time on materials and so on, construction time (fitted into available planned downtime), and backup measures, in the same detail, if you think they are necessary.

2 Show your management the costs—both the probable and the possible extreme cost—which are likely to be incurred in any budget year.

3 Show OSHA the rate at which you expect to make progress in getting people below a 1.0 C/T.

4 Show both your management and OSHA those particular problem areas where no reasonable amount of engineering effort is likely to attain the exposure required by law.

The cost and time estimates you have made at this point need not be conjectural, though they may not be absolutely accurate. Therefore, on the question of cost and time requirements you will want to be generous but realistic. Here's how that works.

Let's say that in a certain room there are several noisy machines and that although it is possible to point to a particular part of one as the main cause of noise in some locations, there are places that need to be quieter where you cannot identify any specific sources. For these locations absoption surely occurs to you as a possible answer. If you can estimate how much absorption would be required to attain your design goal in such a location, you have found a kind of top limit for cost. It is a limit because you will have to do something in the other locations where you know what the source is. And this will help you in the location where you allowed for absorption.

On the other hand, don't get locked into absorption just because it occurred to you. Even if you don't solve the problem by getting at the original noise sources, it might be possible to put a machine operator in a booth, for example.

As you mentally try out these possibilities and estimate what they would cost, you will find a realistic range. Do not count on clever and inexpensive solutions at this point—you can look into these when you begin to work on the methods that will be used.

GETTING UNDER WAY—AT LAST

Now we come to what everybody thinks is industrial noise control. Don't be discouraged with all the work that really should be done before you can reach this point. With some experience behind you, all the foregoing occupies only about 10% of the total time for completion of the program.

With or without some preliminary measurements (we talked about "snooping" and using a stethoscope) you will want to make a list of noise sources at each of the locations. Then start to think, and note, how the noise gets from each source to the operator. A diagram (Figure 88) will evolve.

Sometimes the source-path-receiver diagram will be much more complex than this. If it is much simpler, you probably won't bother to make one because the source and dominant path will be obvious to you.

Now look at each block of your diagram and do your best to estimate how important it is in the whole picture and what might be done to reduce the noise or structureborne vibration.

For example, which is more important for the electric motor, the air noise or the vibration? Is either of them serious in the whole context? Does the motor need an overhaul? Would a motor silencer help?

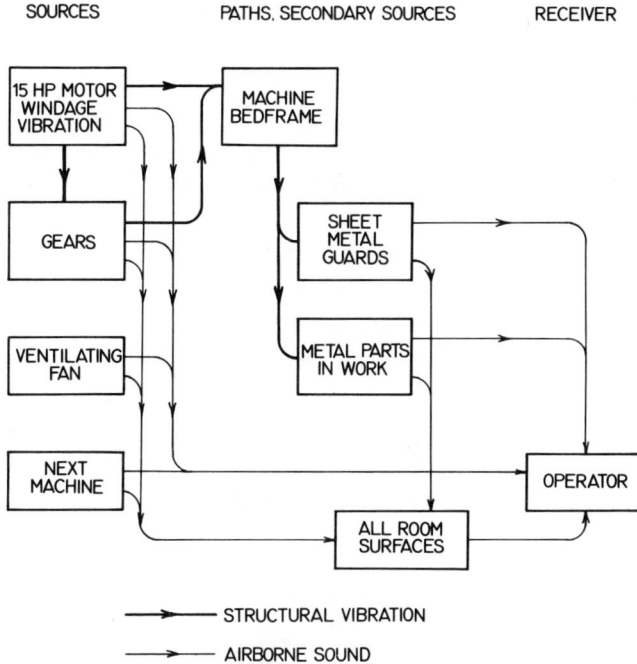

Figure 88 In a complex situation where the noise problem is not easy to diagnose, a source-path-receiver diagram may help organize your thinking. How important is each source or secondary source? Are some paths more efficient than others? Can some paths be broken, blocked, or made less efficient quite easily?

When you have gone through the whole diagram you might have a list that looks like this:

Fifteen-Horsepower Electric Motor

1 Is in good repair and is not a major source of noise or vibration. No improvement of noise level at operator position available through work on motor, or a silencer for it.

Gears

1 Major source of air noise and vibration in machine bedframe. Gears are worn. Replacement cost in the range $5000 to $10,000 and will require 6 weeks downtime for overhaul. What will the life of the machine be with and without repair work? Should the machine be replaced?

2 Enclosing open gear train is not practical. This would not solve vibration problem anyway.

3 Could some gears be replaced selectively? Fiber or nylon gears? Check

possibilities and costs with manufacturer. Tentative cost is in the range of $3000 to $5000.

Next Machine

1   Will have to be worked on for noise at other location. Note that when it is down, level at this machine's operator position drops 2 dBA, not enough to meet 6 dBA design goal.

Ventilating Fan

1   Found this out of balance. Effect at this location is small (0 to 1 dBA by turning it off) but asked maintenance to rebalance anyway because it will get worse otherwise.

Path From Motor and Gear to Bedframe

1   No way to break path or insert vibration isolation.

Bedframe

1   Can't think of any possibilities.

Room Surfaces

1   Reflected noise from floor is significant even though operator is in the direct field of the machine. Absorption on floor is not possible. Partial barrier could be used to break the path, but production and maintenance say they couldn't live with it.
2   Reflected noise from walls and ceiling is not serious and absorption will not do much. (Note absorption may be required for other locations. The maximum effect here will be about 1 dBA).

Sheet Metal Guards

1   A major source. May be resonant with gear frequencies. Complex shape and fit would make it hard to replace with other material. Could we change the guarding system? Maybe use expanded metal fence and leave guards off? The effect might be 4 to 6 or more dBA. See if we can remove guards for a trial run. Cost? $2000 maximum for fence. Will production and maintenance be satisfied with this idea?
2   Can guards be isolated with soft gasket material? The effect will be smaller, but might be worthwhile. Cheap!
3   Line the guards with damping material or maybe combination damping-and-absorber to lower airborne noise from gears, too. In combination with isolating gaskets this might achieve the 6 dBA we need. Cost: $800?
4   Stiffen guards by welding on angle iron to change resonance frequency. Ef-

fect? If we do this it will have to come before doing anything else to the guards.

Metal Parts in Work

1 Principal noise when they fall into hopper. Line hopper with rubber. Effect small (0 to maybe 2 dBA) but this is annoying and cure will only cost $75. Have recommended it.

All Air Paths to Operator

1 Could put the operator in a partial enclosure. Cost estimated as $1200, in place, and about 10 dBA noise reduction is available. Production department is worried about the effect on other operators—everyone may want one.
2 Only place barriers are acceptable to production department is on upper half of machine. They would be no help there.

Now this is all *hypothetical,* you realize, but see how it pins down the possibilities. In the first place it rules out things that might have been considered—or actually done (like an acoustical ceiling)—without helping. It has also turned up information about the gears which indicates that overhaul or replacement of the machine deserves to be considered both from the noise standpoint and in order to assure continued production.

Don't be surprised if you are the first person to spot equipment on the verge of failure. It happens all the time in noise control work. People who work with a machine every day are used to it wheezing. You are in the position of taking a fresh, analytical look. (Notice that *you*, and not the maintenance department, picked up that out-of-balance ventilating fan, too.)

It may look to us, concerned with the noise problem only, that work on the guards for less than $1000 is the best alternative. The partial booth for the operator is not far behind. It costs a little more, but it is sure to solve the 6 dBA design goal problem.

How will it look to the production manager and higher management? Different, perhaps. They may elect to have the machine overhauled for $8000 or even buy a new one for much more. Whatever they decide to do, you can be satisfied with the analysis you have made if it comes out like this one.

HOW DID WE DO THAT?

The guards and gears were the culprits on this machine. How did you learn that? Can you prove it, to your own satisfaction, at least? In a case like this your stethoscope is a real friend. It alerted you to gear noise at the bearings of the gear shafts and at the gearbox. It found a noise that sounded quite similar at the guards. Sure enough, when you walked back to the operator position you could hear the same tones. Since the operator position was only 6 to 10 ft from the gears and guards, the only paths worth worrying about were the direct air path and the air path with one bounce on the floor. (If the operator had been farther away, you would have tried cupping your hands behind your ears to see whether much of that tone was bouncing off the ceiling or walls.)

If your meter was a survey or octave band type you might have compared readings (either A-weighed or in the peak octave) at the gears, the guards, and the operator locations. Suppose you did and found:

Meter Location	Level (dB)
6 in. from open gears, typical	99 to 103, **101**
6 in. from gearbox, worst found	**96**
6 in. from sheet metal guard, typical	102 to 106, **105**
Operator location	95 to 98, **97**

Doesn't that tell you something?

If you had narrower band analyzers (they can be rented, if you think it worthwhile) you might take the analysis further. You would pick out the frequencies (or frequency bands, if you were using one-third-octave equipment) where the noise peaks occurred. Then, using a print of the gearbox or data you could take on open gears when the machine was down, you could calculate out the gear clash frequencies. You would be looking for a match between calculated and observed values.

It doesn't always work out so neatly, but you will sometimes pull a real Sherlock Holmes. In one case a consultant phoned a plant he had worked at to warn them that a particular gear pair in a four-shaft gearbox was in bad shape based on this sort of analysis. About 2 weeks later, that gearbox packed up. The pinion of that pair was stripped.

On open gears (or other rotating or reciprocating units) you can sometimes see exactly what the problem is by using a precision stroboscope/tachometer. Before you rent one, find out if the maintenance department owns one. If not, be sure to show the maintenace superintendent how it works . . . he may buy one. By synchronizing the flashes of this instrument with the machine's cycle, you can freeze the motion, and if you cannot (assuming the strobe is in good repair), you have at least learned that there is a lot of chatter in the gears.

If you don't have access to such fancy equipment, don't despair. You can do a great deal with your ears alone. Comparing the tone you find in the metal guards, for example, with that from the gears, the motor, and a fan that might be part of the machine should let you reach the conclusion that it was not the fan. If you heard the same tone in the motor, gears, and guards you would have a small problem. But you would quickly reason that the guards were a less likely original source than the gears and motor. By the same process, you would correctly come back to the gears rather than the motor.

If you have a musical bent or happen to be one of the rare people with absolute pitch, make use of it. A pitchpipe doesn't make enough noise to be heard on the job. You can, however, sing the lowest note you can reach. Then work your way up by octaves, if you have to, or singing the scale until you can match the tone the machine makes. Jot down the musical note you found and look up the frequency later.

SOME HELP WITH THE DETECTIVE WORK

Sometimes looking at the obvious can be helpful. There aren't many ways noise is created. Sometimes when you have reasoned your way back to the cause of noise an effective, inexpensive way of preventing it will occur to you—even though this doesn't happen often. It is such a good idea that it doesn't have to happen often to be a useful addition to your bag of tricks.

Impact

The impact of rigid solid materials causes noise. Drop a part on the floor and—sure enough—bang! But sloppy spur gears also produce an impact as each driving gear tooth hits its partner in the driven gear. One useful idea, therefore, is to substitute helical, herringbone, or spiral bevel gears. In the best of these the next pair of teeth is picking up the load before the first pair have finished transmitting their share of the power. And for impacts that can be avoided or cushioned, dramatic improvements are possible here.

Turbulence

In flowing fluids, gaseous or liquid, turbulence generates noise. Turbulence is often unavoidable and you will have to cope with the noise produced by add-on noise controls. Velocity does not always have to be so high or the section so small and here you can sometimes make a dramatic change. The sound power produced in some of these situations involving gases increases as the eighth power of velocity. Under these rules a little can accomplish a lot. Pulses in gas streams produced by a repetitive sequence of events produce tones and their harmonics. Your car engine is one culprit here. Fans and some positive displacement blowers are common sources of this type of noise in the plant. Sometimes you can get the same performance with a multistage centrifugal blower at not so high a price in pulses and tones.

Combustion

Combustion noise—especially when you are burning gas or atomized oil—is something like the turbulent boundary situation described in Chapter 10 on jets. With combustion noise there is also a way in which the inherent noise mechansim can act to direct a natural amplifier. A lot of energy is available because in a combustion process the combustion product gases are heated, and thus they expand very rapidly. Usually little can be done about the low frequency rumble of a big furnace or the hiss of an oxy-acetylene torch or even the mid-frequency braying of a Meeker burner, because the intense turbulent mixing is exactly what is desired for efficient combustion—and that probably takes precedence over a design that makes less noise. The cost of quiet burners and furnaces is too high.

Friction

Friction may be the most known and least understood phenomenon associated with twentieth century technology. Moreover, in general, whatever you do to reduce it will reduce noise *and* make you a hero. Repetitive "stick-slip" cycles are one mechanism often suggested for what is actually happening when frictional forces assert their drag. If this mechanism is really what usually happens, it must be producing noise perforce. Now although no deep insight into the mechanisms of friction is offered here, it is noted that adding a premium lubricant like molybdenum disulfide to the oil of a noisy gearbox usually produces a noise reduction of 2 or 3 dB and has achieved 8 dB. Some of the noise produced by gears is surely impact noise (especially if they are spur gears) but evidently a major part of the noise is sometimes rubbing and sliding friction.

Imbalance

Imbalances of rotating equipment or dynamic forces generated by reciprocating equipment are so obvious a cause of noise that they need hardly be called to attention. In the case of rotational imbalance, at least, the solution is equally obvious. For the cases where that imbalance is an intentional part of the design the question should arise: Is this the best design?

Electromechanical

Electromechanical effects can be causes of noise. An electric motor with a wound armature has wires that are laid into slots in the iron core of the armature. Subjected to the combination of magnetic flux and a current running through them, these wires supply the force that makes the armature go around. What if the wires are loose in their slots? Then each time they exert the force they also slap the side of the slot—and that can be often enough to broadcast an intense pure tone.

This is not the only electromechanical noise source, however. Laminated transformer cores manage to slap together twice in each electrical cycle despite the transformer manufacturer's best effort to build rigid, gap-free cores. There are other, similar cases.

In the case of screaming wound-armature motors, the noise is called *slot* noise. The simplest and least costly solution may be to have the armature re-wound with the windings bedded securely in epoxy.

Air-pocketing

Air-pocketing may be likened to *total break free.* Both phenomena may seem obscure, but they can be causes of noise. For air-pocketing, imagine a machine that prints and die cuts sheets of corrugated board to make cartons. The sheets are stacked at the feed end and withdrawn from the bottom of the stack. Each new sheet, in falling onto the table, traps air beneath it and that air may have to travel a few feet to find an edge where it can escape. If the process takes place in a hurry (and of course the plant manager has an interest in making it happen that way), the escaping air will be traveling at impressive speed as it finally gets away at the edge. This is a source of noise.

Total break free is what happens when you do the most logical thing and design a blanking press to have all cutting and supporting edges true and level and perpendicular to the cutting stroke. The blank breaks free along all edges at the same instant. A lot of energy that has been stored in the stock—both the blank and the surrounding web—by elastic deformation is released at the instant the blank snaps free. Some tool and die makers will tell you, and unless you know as much as they do you must take them seriously, that it is not possible to grind a slight angle into the cutting tool or the support die and still produce a true and flat blank. On the other hand, in some blanking operations it has been done with acceptable results. One important result for you is that because the cut is produced by smooth shearing there is little energy stored in the metal. The blank does not snap out, it drops out. Such presses still produce noise because they move quickly and lose some of their power to wasteful (noise producing) processes. However, they do not employ a process that selectively favors noise production.

FINISHING UP

After management has made a decision on what corrective measures to take, you will probably want to prepare sketches of what is needed. You will specify materials and show any construction details that are essential to the acoustical performance.

Carefully consider all the nonacoustical requirements that can cause acoustical trouble. Have you provided access for maintenance? Will glass fiber shed into a sensitive product? If foam is used, will it be a fire hazard? Are you restricting or slowing down any activities that must be carried out at top speed in emergencies? And so forth.

Consultants usually take the job this far and then turn the shop drawing part of the project over to the client's engineering department or to the contractor selected by the client. They always suggest that they review the drawings and arrangements in these cases. So should you if somebody else is doing the final design. Your reputation in your company depends on it, and also you may find mistakes that would be costly.

THE STICKY QUESTION OF HEARING LOSS PREVENTION

The question of the role of noise exposure in hearing loss is not a nice, clean-cut issue that can be raised and disposed of. You should not take any authority's word as final. Not only is the jury still out, the evidence has not all been presented. In fact, the detectives are still working on (and puzzled by) the case. Let's sidestep the political/philosophical questions of why and how OSHA and workmen's compensation laws get written without some well-established basis. They do get written. This concern for "the quality of life" has put many important forces in motion. One result is that you bought this book!

First, some facts about compensation for noise-induced hearing loss are:

1 Typical awards for noise-induced hearing loss under workmen compensation laws during the past 10 years have increased from about $1000 to more than $5000.

2 Some studies show that about one-quarter of the people working in industry may be able to support such a claim at the time they retire.

3 The number of claims that are made each year are far, far less than this. One explanation offered is that if an employee has lost a finger in an industrial accident, you can see that he has by looking. Perhaps human vanity keeps people from admitting (even to themselves) that they don't hear so well. Labor unions are beginning to point out the availability of hearing loss compensation to retiring workers, however, so this may change.

4 In the case of Federal government employees the number of claims for compensation for noise-induced hearing loss has risen faster than any other sort of claim in the last few years. There are 7000 pending cases with claims totalling $28 million.

5 In a recent case, two retiring employees were awarded $15,000 each for noise-induced hearing loss. The trend is toward more compensation.

6 In California, a recently passed law allows for "psychic trauma" induced by noise exposure. (Won't *that* be a lawyer's field day.)

7 Under a directive to field inspectors for OSHA in April 1979 an "effective hearing conservation program" was defined. It had been required under the 1971 law but never before precisely defined. The new directive said that an effective hearing conservation program required "periodic" hearing tests. (These had already been required annually in Kentucky, Pennsylvania, and Wisconsin.) Does "periodic" mean "annual"? OSHA has not committed itself on that question in print yet but some responsible OSHA people will

tell you "periodic" means "at least annual." More frequent tests would be better, they say, if you notice your employees' hearing deteriorating.*

Some other classic cases argue that noise-induced hearing loss is real and predictable. The first to surface was "boilermakers' deafness" back in the 1800s. Then there's the "bomber's notch," severe loss of hearing in the 6000 to 8000 Hz range experienced by the pilots who "flew the hump" in Burma during World War II. It was shown, after the fact, that the noise levels in the planes they flew (sometimes for 16 hrs a day) were the cause.

Another example is the case of caulkers—they drove oakum into the seams of wooden ships, using a long-headed hammer with a split tail like a tuning fork. To tell when the oakum had reached a critical tightness in the seam, they listened for a change in the intensity of the tone that long split tail made. When tested, they all showed hearing loss in the bands covering that tone and an octave or a little more above it.

The first learned report on the subject, the CHABA (Committee on Hearing and Bioacoustics) report, was needed because of the hearing difficulties of tank crews in World War II. (CHABA's findings, incidentally, formed the basis of the 1969 noise amendment to the Walsh-Healey Act, the immediate precursor of the OSHA law in effect today.)

There are even scholarly papers that allude to audiograms made on members of remote African tribes where noise has never been much of a problem. The old men of these tribes hear as well as young people in Western culture.

This is not a critical evaluation. With an objective, open mind, listen, now to the other side of the story:

1   There are well-conducted studies that show that the difference in hearing loss acquired in an 82 dBA environment is not discernible from that acquired in a 92 dBA environment.

2   In one consulting company's review of hearing tests—and this is not a conclusive piece of evidence yet because they have not researched the background causes—it is often found that there is substantial hearing loss among people exposed to about 0.25 of the permitted daily noise dose on their jobs, and sometimes no hearing loss for a significant fraction of people regularly exposed to 1.0 or more of the OSHA figure. Moreover, truckdrivers and maintenance men, as a group, regardless of the C/T figure in their daily job show significant hearing loss.

3   Some studies have shown that in noisy working environments about one-third of the workers who regularly wore hearing protection (typical noise

---

*As this book was in final preparation in January 1981, OSHA issued a new regulation covering this subject. The date on which it will take effect had not been settled as the book went to press. It confirms the need for annual tests and exhaustively defines hearing conservation program requirements.

reduction estimated at something like 20 to 25 dBA) fared exactly the same as those who regularly did not.

4   Levels in discotheques are often 112 dBA and higher and pure tones (supposedly more damaging than broadband noise) dominate. Yet study after study has shown no permanent hearing loss from this exposure. In fact, one much-studied group of professional rock musicians who perform night after night does not show hearing loss either.

If your goal is to do "the good thing" and prevent hearing loss from noise exposure, or even the selfish thing of minimizing workmen's compensation claims, you will require great knowledge and great skill. There is no clear answer or advice for you in today's literature or wisdom.

What course is open to you or to the considered opinion of your management, then? Surely you must make a realistic effort to meet the goals that OSHA has set for you. That is a kind of traditional wisdom that no study of the data or learned comment can ever overcome. It is the distillation, in the form of enacted law, of the wisdom of this age. You flout it at your peril.

What else can you do? You and your company can recognize that we do not yet know the answers and you can gather the best evidence that it is now possible to gather: you can make tests, at least annually, of the hearing of your people. It can be said that you did the best you could.

GLOSSARY
Terms Used by OSHA

Adminstrative controls   Any procedure that significantly limits daily exposure by control of the work schedule is an administrative control. The use of personal hearing protection *is not* a means of administrative control.

Compliance program   *(abatement plan)*   In the implementation of controls, a step-by-step program for correction will enable the industrial hygienist to monitor abatement progress. It will also provide a means for the employer to make decisions and set schedules and for the employee to understand the changes being made. Large projects will require a detailed program and the industrial hygienist will decide whether the plan submitted for such cases is sufficiently detailed.

Determination of noncompliance   As a minimum, sampling will be conducted for the time needed to establish whether a violation exists. When a reading of 132% is obtained on a dosimeter, noncompliance is established.

Dosimeters   When sound level meter readings invalidate dosimeter readings a citation will not be issued based on the dosimeter results.

Engineering controls   These are physical means that reduce the noise causing the exposure. They include substitution of manufacturing equipment or process, isolation brought about by barriers, enclosures, and the like, or modification of the equipment including the addition of materials like absorbers and damping materials.

Equivalent A-weighted level   An equivalent level can be found from a dosimeter reading by using

$$L_A = 90 + 16.61 \log (D/12.5T)$$

where L_A = the equivalent A-weighted level (dBA)
  D = the dose read from the dosimeter (%)
  T = the sampling time (h)

(Generally OSHA will report this to the nearest tenth of a dBA.)

Feasibility   The existence of general technical knowledge of materials or methods that are available and suitable or adaptable to the circumstances and that can be applied with a reasonable expectation that the noise level will be reduced.

Impulse noise   Impulse noise includes noise caused by impact of two objects. it also includes any noise where the level rises to a maximum in not more than 35 msec and not more than ½ sec passes before the level falls to at least 30 dBA below the peak. If such impulses recur more often than once a second, they will be considered continuous noise.

Words Used in This Book

Absorption, *A*, in sabins   The capacity of a surface to absorb sound power impinging on it without reflecting that power. In a room or other space bounded by several surfaces the total absorption is found by

$$A = S_1 \alpha_1 + S_2 \alpha_2 + S_3 \alpha_3 + \ldots$$

where A = the total absorption at some frequcney (sabins)
   S = the surface area of parts 1, 2, 3, etc. (ft^2)
   α = the absorption coefficient of these parts at the frequency of interest (no units)

Absorption coefficient, α, no units   That fraction of the sound power that is not reflected when sound impinges on a surface once. The coefficient is usually a function of frequency. Although it is also usually a function of the angle of incidence, absorption data not otherwise specified can be assumed to be based on random incidence. (Coefficients found in "tube tests" represent normal—90°—incidence and are not reliably related to those found in reverberation room tests.)

Attenuation, *A*, in decibels   Often subscripted to distinguish it from *absorption*. Thus A_{air} would represent an attenuation (or noise reduction) caused by passage of sound through the air. Attenuation is often used to describe the noise reduction caused by several mechanisms acting at once, for example, a wall that is penetrated by a muffler. Attenuation is also sometimes given in units of decibels per 1000 ft or some other descriptor.

Decibel, dB, no units (but *always* referenced)   A division of a logarithmic scale used to express, as a level difference, the ratio of two like quantities proportional to power or energy. This ratio is expressed in decibels by multiplying its common logarithm by ten.

A change in level of 1 dB can only be sensed by immediate comparison and critical listening (e.g., switching a well-controlled tone back and forth from 80 to 81 dB). Three decibel changes are needed to produce a consensus judgement that the noise is really louder or softer. It is conventional to say that a 10 dB change produces a subjective impression of twice (or half) as loud, and 20 dB produces an impression of four times (or one-quarter) as loud. Fletcher and Munson's data say otherwise. So will your ears, if you listen critically.

Perhaps the decibel itself, the central measure of acoustics, captures something of the bewildering uncertainty of the whole subject. Twice as loud? Is it 10 dB (late twentieth century wisdom), 6 dB (in vogue in the 1930s with the tenuous logic of 2^2 or $\sqrt{2}$ factors of power), 5 dB (with no basis evident—*except* that it is the basis of the current OSHA law), 3 dB (so satisfying because it makes sense in power, but so frightening because there are people who want to rewrite the OSHA laws on this basis), or ...?

Frequency, F, in hertz   The rate at which a cycle of sound or vibration is repeated.

$$F = (cycles)/t$$

where F = the frequency in hertz (or cycles per second)
$\quad\quad t$ = 1 sec

$$F = c/\lambda$$

where F = frequency (Hz)
$\quad\quad c$ = the speed of propagation of sound (vibration) in the medium (ft/ sec)
$\quad\quad \lambda$ = the wavelength (ft)

Insertion loss, IL, in decibels (but see the following notes)   The reduction in level caused by inserting some element in an acoustical system. The use of decibels implies a reference power. The term "insertion loss" implies that the reference power was that of the system before insertion of the element. This is why IL can be handled with simple arithmetic rather than conversion to power and so forth.

Intensity level, IL, in decibels   A measure of the sound power impinging on a surface such as a microphone diaphragm or the eardrum. For the temperatures and ambient barometric pressures normally encountered in industrial noise control work, the IL can be considered to be the same as the sound pressure level.

$$IL\,(\cong L_p) = 10 \log\,(I/I_o)$$

where   IL = the intensity level (dB)
$\quad\quad L_p$ = the sound pressure level (dB)
$\quad\quad I$ = the power intensity of the impinging sound (W/m^2)
$\quad\quad I_o$ = the reference power intensity (10^{-12} W/m^2)*

*On an area basis this is equivalent to 10^{-16} W per square centimeter. A square centimeter fairly approximates the size of the eardrum. Thus 0 dB in IL or L_p, based on this standard reference, is also close to the threshold of hearing, for some frequencies, of people whose hearing acuity is normal.

Level   A ratio, expressed in decibels, of two power like quantities—one of which is the reference.

Noise reduction, NR, in decibels   The decrease in sound pressure level brought about in any way. Noise reduction can be achieved by slowing down a machine, for example, or replacing worn parts, or modifying its design. Where NR is achieved by noise control measures external to the source, some of the common causes are.

Absorption

$$\text{NR} = 10 \log (A_f/A_s)$$

where   NR = noise reduction (dB)
A_f = the final absorption (sabins)
A_s = absorption to start (sabins)

Barriers

$$\text{NR} = 20 \log (\sqrt{2\pi N}/\tanh \sqrt{2\pi N}) + 5$$

where   NR = noise reduction (dB)
$N = (A + B - d)(F/565)$ and $A + B$ is the shortest path length with the barrier in place (ft)
d = the shortest path before barrier (ft)
F = frequency (Hz)

(Note that this equation applies to one edge of a barrier only. NR afforded by the other edges, if these exist, must also be considered, lowering the effective NR of the barrier. Indoors or where other significant reflecting surfaces are present, they will also lower the NR achievable.)

Distance

NR = $20 \log (d/d_o)$ for a point source   and intermediate coef-
NR = $10 \log (d/d_o)$ for a line source   ficients are sometimes
NR = $0 \log (d/d_o)$ for a large area source   used in mixed cases

where   NR = noise reduction (dB)
d = some distance of interest (any units)
d_o = the distance for which the L_p is known (same units)

(Note that moving so far from a large area source that it begins to appear as a line or point will increase the coefficient—though often the distance required to do this makes the effect an unimportant one.)

Excess Attenuation

Empirical data exist that permit NR to be assessed for large volumes of air of known temperature and moisture content, trees and other vegetation, and so forth. No simple mathematical expression predicts these effects.

Walls

$$NR = TL + 10 \log (A/S)$$

where NR = noise reduction $\left.\right\}$ at the frequency
    TL = transmission loss $\left.\right\}$ of interest (dB)
    A = absorption on the receiving side (sabins)
    S = the surface area of the wall (ft^2)

Noise reduction coefficient, NRC, no units A conventionalized average of the absorption coefficient to produce a single number rating of absorptive materials useful to architects and often misleading in industrial noise control.

$$NRC = (\alpha_{250} + \alpha_{500} + \alpha_{1000} + \alpha_{2000})/4 \text{ to the nearest } 0.05$$

where NRC = the noise reduction coefficient (no units)
    α = the absorption coefficient for the frequency subscripted (no units)

Resonance A phenomemon, not a quantity. Thus it has neither a symbol nor units. Resonance occurs in a system when two opposing reactive forces equal each other (a spring-restoring force and a kinetic force of a mass in motion, for one example). If there is little frictional loss (damping) and if the driving frequency is close to the frequency at which the two reactive forces are equal, the system will "ring" or resonate. The critical frequencies for some simple systems are:

Mass Vertically Supported By a Spring

$$F = \frac{1}{2\pi} \sqrt{\frac{\text{stiffness}}{\text{mass}}}$$

Long Air Column (Pipe) Open at Both Ends

$$F = \frac{1}{2L} \sqrt{\frac{\gamma \times \text{pressure}}{\text{density}}}$$

where γ = the ratio of specific heats
    L = the length of the column in consistent units

(In this system prominent harmonics can be expected at every integer multiple of the fundamental.)

Long Air Column (Pipe) Open at One End

$$F = \frac{1}{4L} \sqrt{\frac{\gamma \times \text{pressure}}{\text{density}}}$$

where γ = the ratio of specific heats
           L = the length of the column in consistent units

(In this system prominent harmonics can be expected at every *odd* integer multiple of the fundamental.)

Sound power level, L_w, in decibels   A measure of the total sound power being radiated by a source.

$$L_w = 10 \log (W/W_o)$$

where L_w = the sound power level (dB)
           W = the total power radiated as sound by the source (W)
           W_o = the standard reference power* of 10^{-12} (W)

*Some older literature is referenced to 10^{-13} W. For a point source, under the old 10^{-13} W standard, the L_p at 1 ft would be the same as the L_w. Under the new standard reference L_p equals L_w at 1 meter for a nondirectional source.

Sound pressure level, L_p, in decibels   A measure of sound power available from the air at the location where the measurement is made. It can be seen to be similar to intensity level and for all practical purposes it is the same. The difference between the two quantities arises because L_p is actually measured by sensing pressure changes in the air caused by the sound. The process of transferring the power to the microphone (or eardrum) is less efficient when the air density is low and more efficient when it is high. The variation between IL and L_p as measured in air remains less than 1 dB for altitudes from 10,000 ft above to 1000 ft below sea level, and temperatures from well below 0 to over 200°F. The *useful* distinction between the two quantities is that intensity level can *only be calculated* and that L_p *can be measured* with workable reliability and can also be calculated. By definition,

$$L_p = 20 \log (p/p_o)$$

where L_p = the sound pressure level (dB)
           p = the dynamic pressure variation caused by the sound ⎫(in consis-
           p_o = the reference pressure (20 μPa/m^2)   ⎭tent units)

However, while this equation serves to define the quantity L_p, it is all but useless in noise control work.

The working equation relating L_p to L_w is given, with explanation, in the section "Let's Level" of Chapter 3.

Sound Transmission Class, STC, no units (decibels implied)   The single number index of a wall's merit in transmission loss. It was instituted to be helpful to architects. In industrial noise control it has the single virtue that it means a test has been made. Thus the actual TL behavior of the wall may be available if you can get the test data.

Transmissibility, various symbols, no units   A ratio of force (or displacement) transmitted through a vibration isolator to the force (or displacement) applied to it.

Transmission Loss, TL, in decibels   The reduction in sound power level brought about by a wall. It is usually quite close to the reduction in sound pressure level also (see following equation). Like insertion loss, TL can be handled with simple arithmetic because the reference on which its decibels are reported is the power at the measurement point *before* the wall was in place.

$$TL = NR - 10 \log (A/S)$$

where   TL = the transmission loss for each frequency of interest (dB)
        NR = the measured noise reduction for each frequency (dB)
         A = the absorption for each frequency (sabins)
         S = the area of the wall (ft^2)

Transmissivity, τ, no units   A ratio related to TL that has nothing to do with transmissibility. It has not appeared in this book but is a useful way of handling some calculations involving leaky walls or walls of mixed TL. It will be recognized as one of the ingredients of the equation for estimating flanking (Chapter 9). In the equation

$$\tau = 10^{-(TL/10)}$$

or, alternatively,

$$TL = 10 \log [1/\tau]$$

Wavelength, λ, feet   The distance between successive compression fronts of a sound wave in air. (This is the usual meaning in this book. It is, of course, sometimes useful to work with media other than air and, for solid media, sometimes other sorts of waves, *eg*., bending.)

$$\lambda = c/F$$

where λ = the wavelength (ft)
         c = the speed of sound (ft/sec)
         F = the frequency (sec^{-1})

APPENDIX

Table 22 *Hyperbolic Tangents (tanh x)*

x	tanh x	x	tanh x	x	tanh x
0.100	0.100	1.10	0.800	2.2	0.976
0.150	0.149	1.15	0.818	2.3	0.980
0.200	0.197	1.20	0.834	2.4	0.984
0.250	0.245	1.25	0.848	2.5	0.987
0.300	0.291	1.30	0.862	2.6	0.989
0.350	0.336	1.35	0.874	2.7	0.991
0.400	0.380	1.40	0.885	2.8	0.993
0.450	0.422	1.45	0.896	2.9	0.994
0.500	0.462	1.50	0.905	3.0	0.995
0.550	0.501	1.55	0.914	3.2	0.997
0.600	0.537	1.60	0.922	3.4	0.998
0.650	0.572	1.65	0.929	3.6	0.999
0.700	0.604	1.70	0.935	4.0	0.999
0.750	0.635	1.75	0.941	4.5	1.000
0.800	0.664	1.80	0.947	for all	
0.850	0.691	1.85	0.952	higher	
0.900	0.716	1.90	0.956	values	1.000
0.950	0.740	1.95	0.960		
1.000	0.762	2.00	0.964		
1.050	0.782	2.10	0.970		

Table 23 *Target PNC Curves for Spaces Based on Intended Use.*

Use	PNC
Small auditorium, classroom, or meeting room where no amplification will be used	35 or lower
Offices for one or two people, small meeting rooms	30 to 40
Large offices, cafeteria	35 to 45
Drafting room, stenographic pool, lab areas	40 to 50
Maintenance shop, pilot plant, other situations where the people must be able to communicate with each other	60 maximum

INDEX